Total Eclipses of the Sun

Total Eclipses of the Sun

J. B. Zirker

PRINCETON UNIVERSITY PRESS
PRINCETON, NEW JERSEY

Published by Princeton University Press, 41 William Street,
Princeton, New Jersey 08540
In the United Kingdom by Princeton University Press, Chichester, West Sussex

Originally published in hardcover by Van Nostrand Reinhold Company, Inc., 1981. Reprinted by arrangement with Paramount Publishing, 113 Sylvan Avenue, Englewood Cliffs, New Jersey 07632

Library of Congress Cataloging-in-Publication Data

Zirker, Jack B.
Total eclipses of the sun / J. B. Zirker. — Expanded ed.
p. cm. — (Princeton Science Library)
Includes index.
ISBN 0-691-02952-0
1. Solar eclipses. 2. Solar eclipses—1991. I. Title. II. Series.
QB541.Z57 1995
523.7'8—dc20 94-47241

Princeton University Press books are printed on acid-free paper and meet the guidelines for permanence and durability of the Committee on Production Guidelines for Book Longevity of the Council on Library Resources

First paperback printing, for the Princeton Science Library, 1995

Printed in the United States of America

Contents

Preface to the Paperback Edition

In 1984, when this book first went to press, astronomers were preparing to observe the total solar eclipse in Java. As I write this preface, we are preparing to observe yet another eclipse in India. A lot has happened in the decade between these two eclipses. Technology has advanced—most professionals now use solid-state arrays instead of film, as well as video cameras and portable computers. X-ray cameras of unprecedented resolving power, aboard the Japanese satellite "YOKOH" and on American rockets, have been used to observe the corona without an eclipse. New detectors now enable astronomers to make images of the Sun in infrared (heat) waves. Some questions about the Sun have been answered definitely in the past ten years, but others—such as the source of heating of the corona—still remain to challenge us. Amateur interest in eclipses has exploded: hundreds of tourists now travel thousands of miles to the remote places of the world in order to view the next spectacle.

Since 1983, six total solar eclipses have occurred, but rather than describing each of these, I have chosen to concentrate on the most dramatic and productive of them: the eclipse of July 11, 1991. In chapter 12—new for this edition—you will learn how solar astronomers used some of the largest telescopes in the world to observe the eclipsed Sun and the results they obtained. I hope you will experience some of the excitement we felt during this eclipse, and appreciate the huge amount of work that such expeditions require.

This is a good place to acknowledge the continuing support of eclipse science by the National Science Foundation. During the past twenty years the Foundation has been the principle source of funding for eclipse research by U.S. scientists. In addition to grants for equipment and travel, the Foundation has provided logistical support—electrical power, housing, medical services, and the like—for large groups, especially in the Sahara in 1973. I want to thank the program directors in the Astronomy Division for their unstinting support and encouragement, without which much of the research reported here would never have seen the light of day.

Preface

"Sixty seconds to second contact—MARK!" I watch the digital clock in my hand counting down the seconds, racing toward the final climactic moment of the 1980 eclipse. Have I computed the correct time? Will my voice fail in my excitement during the final ten-second countdown? Thirty of my colleagues hunch around their instruments, making last-second adjustments, checking their equipment for the thousandth time. They look calm, even indifferent in the eerie half-light. Why am I sweating? "Forty seconds—MARK!" I bellow. My left hand holds the lens cap for our twelve-inch telescope. All I have to do is take it off at *exactly* the right instant, without jarring the telescope and then replace it at exactly the right instant two minutes and forty seconds later. Ray Smartt and Horst Mauter, my friends from Sacramento Peak Observatory, will do all the rest for our experiment during the eclipse.

We've worked hard for two weeks in the dry heat of an Indian February to erect the massive telescope mount, set up our filter and camera and our few pieces of electronic gear. We've planned and practiced and made hundreds of test exposures. Everything and everyone must function perfectly or the whole effort will be lost.

The air is cool, the sky clear. The crowds of Indian villagers are beginning to cheer and whistle. I tear my eyes from the clock and snatch a quick look at the Sun. Only a thin bright crescent remains on the east side, and it's shrinking at an incredible pace as I watch. Damn! Have I miscalculated? Surely, the crescent won't last another twenty seconds! What time is it? "Twenty seconds—MARK!" I nearly miss the call. It's so easy to become hypnotized by that crescent. The horizon is yellowish or orange, the zenith a pale blue. No stars yet. The spectators are making a terrific racket. Can I be heard? Cameras are beginning to clack and whirr all around me. What time is it?

"Ten — nine — eight — seven — six — five — four — three — two— one—CONTACT!"

I pull off the lens cap. Ray adjusts the pointing of the telescope a hair and says "**GO!**" Our motorized camera begins to whine and snap regularly, as Horst sets the time for increasingly long exposures. I look up. Incredible! It is the eye of God. A perfectly black disk, ringed with bright spiky streamers that stretch out in all directions. A few red prominences. A star or two. This fantastic object, blazing in the surrounding blackness, at mid-morning. What a stunning sight! The Indians applaud, yell, whistle, while the scientists continue to hunch over their dials, muttering prearranged commands to each other.

"Ten seconds to third contact —MARK!" I holler.

"Five—four—three—two—one—CONTACT!"

The eclipse has ended. I replace the cover. Already the light is changing, remarkably bright an instant after totality. It is a second dawn.

"AUTHOR, AUTHOR," somebody yells. Horst, Ray, and I begin to cheer, to laugh, to dance, and pound each other in glee.

"How did it go? Did the stuff work?" Everyone is grinning, chattering—pandemonium.

"That was a good rehearsal. Now the *real* eclipse begins in ten minutes," some smart aleck quips.

The breathtaking totality is over. Now the work begins anew—recording immediate impressions, taking calibration exposures, dismantling the equipment, and repacking. We will develop our films tonight and know the truth. I've seen perhaps ten seconds of the actual eclipse, but, damn! it was worth it. Sometime, I promise myself, I'll come without an experiment, without even a camera, and just watch the whole time.

The Indian eclipse was my fourth. I'd been to Bolivia in 1966, Mexico in 1970, and West Africa in 1973. Compared with some of my colleagues, who have witnessed a dozen eclipses before they were forty-five years old, I'm still a novice. M. Waldmeier, the Swiss astronomer, probably holds the record, but I'd guess it will fall soon.

More and more people are traveling to eclipses every year. Some of them are professional scientists, some are amateurs, but most are plain, ordinary men and women who hunger for exaltation, for a peak experience. Eclipse watching is safer than skydiving, less strenuous than mountain climbing, but more addicting than liquor. It's not inexpensive either, particularly if you want comfort. These days, you can take an ocean cruise to the path of totality, photograph the eclipse, and have a champagne lunch to celebrate. You can fly during the eclipse (in your friend's plane or your own) to avoid the crowds and the clouds. You can take a deluxe tour, with all the arrangements made for you, and with an astronomer as guide. Or you can join the "eclipse-trekkies" who travel cheaply and camp out near the eclipse track.

It doesn't matter how you get there. If you don't experience the splendour of an eclipse at least once, you can't claim you've lived a full life.

I've written this book for eclipse bugs like me. It is not a collection of anecdotes, however, nor a guidebook to eclipse photography. There are such books (I've listed some) but I've tried to present another side of the matter—the science of eclipses. Eclipses began, and remain, as superb, infrequent occasions on which scientists can do difficult experiments. Astronomers have traditionally put up with heat, insects, boredom, disease, and the orneriness of hardware, in order to carry out some critical investigation at eclipse that simply cannot be done otherwise. But so have meteorologists, geophysicists, relativists, aeronomists, and, lately, biologists.

This book, then, is basically a sampler of eclipse experiments, done in the last decade, by all sorts of scientists. I hope to give you a feeling for the contribution that eclipses have made to several kinds of science.

We begin with two chapters of general background. Chapter 1 skims over the history of man's interest in eclipses, initially for the phenomenon itself, and more recently for information about the Sun. In Chapter 2, we reconstruct the causes of the regularities that eclipses display. Chapter 3 begins a series of topical chapters—each devoted to a different scientific discipline or at least a different aspect of a discipline. We will learn how eclipses test ideas on the deceleration of the Moon in its orbit, and how they reveal the extraordinary structure of the Sun's atmosphere and of interplanetary dust. Next we turn to the Earth's atmosphere and review what eclipses can teach us about its chemistry and physics. One of the classical tests of Einstein's general theory of relativity follows next. Then we consider biological experiments at eclipses. Finally, we speculate about the future. Will eclipses continue to draw scientists or have they been superceded by other techniques?

A word of caution is in order. We will look at a few, highly *selected,* aspects of a few sciences, all chosen because they relate to solar eclipses. Please don't imagine that I'm describing the complete development of each science, or even its present status. I've purposely focused on *eclipses* as a tool for scientific investigation, but obviously, many other tools exist, and many other questions exist that cannot be answered by an eclipse experiment. A certain amount of distortion of the full scope of each subject has been unavoidable.

We will need some concepts in physics and astronomy to appreciate the experiments I've described in this book. I've tried to include enough background, in its proper place, to help readers without much training in the sciences. All this without mathematics! Specialists may find these explanations very superficial, and, if so, I invite them to read on. Nonspecialists, of every age and persuasion, may (and I hope *will*) benefit from them. As further aids to the uninitiated, I've compiled a glossary of unfamiliar names, added a few specific notes of explanation and assembled suggestions for further reading.

You can read the book almost in any order you please. Enjoy yourself!

J. B. ZIRKER

Total Eclipses of the Sun

1
Thirty Centuries of Eclipse Watching

A total eclipse of the Sun is a spectacular, awe-inspiring event. Throughout the past, human beings must have stopped in their daily struggle to survive in order to stare in disbelief at the gradual disappearance and reappearance of the Sun. We have no written records for most of this period. We can only guess at the reaction of neolithic peoples by extrapolating from the reaction of "primitive" peoples now. With few exceptions, they view a total eclipse of the Sun with fear, guilt, a desire to propitiate their gods, and afterwards, an enormous sense of relief.

Through most of recorded history, eclipses were viewed as a supernatural manifestation, usually an omen of some impending calamity. The idea that eclipses are linked to human events changed only slowly, and the idea was manipulated by the priestly class for the greater good of the state. Only the ancient Greeks were comfortable enough in their world to begin to view eclipses with detached interest.

A scientific study of eclipses did not begin until the 17th century, and barely got going until the middle of the 19th century. Thus, although men have been watching eclipses for 300 centuries, and have been recording them for 30 centuries, practically everything we know about them today has been learned in one century, and much of that in the last 40 years. This pattern of extraordinary acceleration in the pace of investigation is common throughout the recent history of science.

ANCIENT AND MEDIEVAL ECLIPSES

Throughout all of recorded history, learned men have written down their impressions of total solar eclipses. Scribes, astrologers, court annalists, official historians, and monks have all contributed to a vast pile of semiastronomical records. These records have modern value, not only for the history of science but for the history of ancient civilizations, because a clear record of an eclipse establishes a precise date in the past. A modern historian, trained in ancient languages, can supply an astronomer with the basic facts concerning an eclipse. The astronomer can then calculate the date and time of the eclipse with an uncertainty of only a few minutes if he is given its location and a date that is accurate to within a decade. Between them, the astronomer and the historian can relate the chronology of an ancient people to our modern calendar system.

This method has its pitfalls, however. In many cases, the original reference is extremely vague and more than one eclipse might qualify as a candidate. Moreover, some of the old historians (like Herodotus) embroidered the facts or even invented eclipses simply to stress the importance of a battle or the birth of a king. The modern scholar must decide whether a record is reliable as well as what it says.

R. R. Newton, who combines antiquarian and astronomer in one person, distinguishes three kinds of spurious eclipses:

1. *The Assimilated Eclipse.* A chronicler may shift the date of an eclipse by a year or more, consciously or unconsciously, to relate it to another event.
2. *The Literary Eclipse.* This type appears in a work of pure fiction, and is itself pure fiction, but is taken as real by some overeager reader. The eclipse of Plutarch (A.D. 71) is an example.
3. *The Magical Eclipse.* "Solar eclipses have a remarkable tendency to happen during battles, at the deaths of great personages or at the beginnings of great enterprises." Herodotus, for example, inserts a magical eclipse into history to punctuate the beginning of Xerxes' campaign against Greece.

This tendency of ancient historians to dramatize their work greatly complicates the work of modern scholars. But several investigators have risen to the challenge, examined reams of old documents and attempted to compile a reliable sequence of past eclipses. Table 1.1 compares the judgments of three experts, who have sifted the original evidence for reliable records of "large" (total, annular, or nearly total) eclipses.

What, then, is the oldest reliable record of a total eclipse? You

TABLE 1.1. Three Authors' Lists of Reliable Pretelescopic Eclipses

	Date	Place	Author*
B.C.	2165? 1948?	China	Needham
	1375 May 3	Ugarit	SC, M
	1330 June 14	An-Yang	M
	1131 Sept 30	Gibeon	M
	763 June 15	Nineveh	M, N
	709 July 17	Chu-Fu	M, N, SC
	601 Sept 20	Ying	M, N, SC
	549 June 12	Chu-Fu	N, SC
	442 March 11	China	N
	431 Aug 3	Athens	N
	424 March 21	Athens	N
	392 Aug 14	Chaldonea	N
	382 July 3	China	N
	364 July 13	Thebes	N
	322 Sept 26	Babylon	M
	310 Aug 15	Sicily	N
	300 July 26	China	N
	198 Aug 7	Chang-An	M, SC
	188 July 17	China	N
	181 March 4	Chang-An	N, SC
	147 Nov 10	Chang-An	N
	136 April 15	Babylon	M, N, SC
	89 Sept 29	China	N
	80 Sept 20	China	N
	28 June 19	China	N
	2 Feb 5	China	N
A.D.	2 Nov 23	China	N
	59 April 30	Armenia	N
	65 Dec 6	Kuang Ling	M, N, SC
	120 Jan 18	Lo Yang	M, N, SC
	243 June 5	China	N
	360 Aug 28	China	N
	429 Dec 12	China	N
	484 Jan 14	Athens	M, N
	516 April 18	Nan-Ching	M, N
	522 Jan 10	Nan-Ching	M, SC
	590 Oct 4	Mediterranean	N
	840 May 5	Bergamo	N, SC
	912 June 17	Cordoba	SC
	916 June 17	Cordoba	M
	968 Dec 12	Constantinople	M, N, SC
	975 Aug 10	Kyoto	SC
	1033 June 29	Europe	N
	1079 July 1	Alcobaca	M

(*continued*)

TABLE 1.1 (*cont.*)

Date	Place	Author*
1093 Sept 23	Mideast	N
1124 Aug 11	Novgorod	M, SC
1133 Aug 2	Salzburg	M, N, SC
1140 March 20	Europe	N
1147 Oct 26	Europe	N
1176 April 11	Antioch	SC
1178 Sept 13	Vigeois	M, N, SC
1185 May 1	Europe	N
1187 Sept 4	Europe	N
1191 June 23	W. Asia	N
1207 Feb 28	Europe	N
1221 May 23	Kerulen R.	M, SC
1239 June 3	Mediterranean	M, N, SC
1241 Oct 6	Stade	M, N, SC
1263 Aug 5	Mideast	N
1267 May 25	Constantinople	M, SC
1310 Jan 31	Europe	N
1361 May 5	Mt. Sumelas	M
1406 June 16	Braunschweig	SC
1415 June 7	Prague	SC
1485 March 16	Melk	SC
1560 Aug 21	Coimbra	M, SC
1567 April 9	Rome	M, SC

*M = P. M. Muller
N = R. R. Newton
SC = F. Stephenson
J. Needham wrote *Science and Civilization in China*. (See the "Suggested Reading List" at the end of this chapter).

might imagine that the Egyptians would have been the first to note such an event, since their historical records date back to at least 4000 B.C., but for some reason the Egyptians are not even in the running. As Table 1.1 shows, all the candidates for the oldest eclipse record are Chinese or middle eastern.

The oldest of all dates to the time of the Hsia dynasty (2183 to 1751 B.C.). It appears in Shu-chin (literally, "The Book"), and is associated with the story of two astronomers, Hsi and Ho, who failed to predict a solar eclipse and were punished by decapitation. This story has come down to us through Gaubil's *Treatise on Chinese Astronomy,* which dates from A.D. 1732. To this very day, scholars argue whether the event of this date was an eclipse, and whether in fact the event occurred at all. The problem is that the Shu-chin was burned in a general repression of thought in 223 B.C., and was reconstructed in the 4th century, A.D.

However, P. K. Wang and G. I. Sisco quote a 5th century B.C. book that quotes yet another book written by Confucius, in which the Hsia dynasty event is mentioned unequivocally as a solar eclipse. J. Needham, in his monumental *Science and Civilization in China,* offers dates for this eclipse that run from 2165 to 1948 B.C.

R. R. Newton concludes that this eclipse was a myth because there are no words in the record that explicitly relate to a solar eclipse. Newton is extremely skeptical of ancient Chinese court annals in general. They contain over a thousand eclipses, but most of them cannot be verified by modern calculations. Were they due to clerical errors, faulty locations, or simple invention? We may never know.

The next most venerable record dates to 1375 B.C. "The Sun was put to shame and went down in daytime," says the historian, recording an event that was seen in Ugarit, an ancient city north of Latakia, in Syria. The date was not recorded, but given as "the day of the new moon in the month of Hiyar," which corresponds to April or May. A computer investigation by J. Sawyer and F. Stephenson yielded May 3, 1375 B.C. as the only acceptable date.

During the Shang dynasty in China, (1766 to 1123 B.C.), events of astrological importance were inscribed on oracle bones. According to Needham:

> Liu Chao-yang has suggested that oracle bones of the second millenium may contain the first recorded observations of the solar corona during an eclipse. The dates of bone fragments studied by Liu must be either 1353, 1307, 1302, or 1281 B.C.

The eclipse of June 15, 763 B.C. appears in the Assyrian Chronicles. The inscription reads "Insurrection in the city of Assur. In the month Sivan, the Sun was eclipsed." This eclipse plays a crucial role in establishing historical dates. Together with lists of kings and reigns, it establishes Assyrian chronology on our own modern calendar.

Perhaps the most famous of all antique eclipses was that of Thales, in 585 B.C. Herodotus in his *History* relates that the Lydians and the Medes were battling in the sixth year of their war when

> . . . day was on a sudden changed into night. This event had been foretold by Thales, the Milesian, who forewarned the Ionians of it, fixing for it the very year in which it actually took place.

Most of the controversy about this eclipse concerns not its date or its place (the neighborhood of the river Halys), but whether Thales knew enough astronomy to predict a total eclipse. Some scholars incline to the view that he was aware of the Babylonians' use of the saros cycle of 18 years in which eclipses repeat, and simply extrapolated from the eclipse of May 18, 603 B.C. However, other specialists believe that even the Babylonians of Thales' time were unable to predict solar eclipses, using the saros, although they might have predicted lunar eclipses.

The Chinese lagged the Greeks in their understanding of eclipses. Shih Shen, a Chinese astronomer of the 4th century B.C. realized that the Moon played some role in eclipses, but thought that its Yin influence overcame the Yang influence of the Sun. Liu Hsiang first proposed the modern explanation around 20 B.C.: "When the Sun is eclipsed, it is because the Moon hides him as she moves on her way."

Theon was an Alexandrian astronomer of the 4th century A.D. He was the first to record, not only the date of an eclipse (in A.D. 365) but also the times of its beginning, middle, and end. Theon must have used a clepsydra, a water clock, and measured time in fifths or sixths of an hour. High precision for those days! As we shall see in Chapter 3, Theon's careful record still counts in recent studies of the Moon's motion.

Hipparchus, perhaps the most eminent astronomer in Greek history, used two observations of the eclipse of 130 B.C. to estimate the Moon's distance from the Earth. The eclipse was total at the Hellespont but only 80% of total in Alexandria. Hipparchus already knew the difference in latitude between these places, from simultaneous observations of the same star. Thus, in effect, he knew their linear separation, in units of the Earth's radius. The eclipse observations now gave him the angular displacement of the Moon (its parallax, in astronomical parlance) as seen from the two locations. Trigonometry, which he was the first to apply to astronomy, then yielded the Moon's distance. He estimated it at 62 to 74 Earth radii. The current value is 60.27.

Evaluating the credentials of ancient eclipses is a harmless indoor sport, played by a small number of ardent specialists with, to be sure, the most serious of motives. It is fun to read how one expert refutes the arguments of another, using nothing more than his presumably better judgement of the cryptic historical records. For example, J. K. Fotheringham, in a key work on the Moon's motion, included in his 1920 analysis the eclipse of Archilochus, a Greek poet who wrote:

> Nothing there is beyond hope, nothing that can be sworn impossible, nothing wonderful, since Zeus, father of the Olympians, made night from mid-day, hiding the light of the shining Sun, and sore fear came upon men.

This presumed reference to an eclipse was identified by T. Oppolzer in 1882, using eclipse calculations and the fact that Archilochus divided his time between the islands of Paros and Thasos. However, that arch-skeptic, R. Newton, dismissed this passage as purely literary, after he found no fewer than *five* eclipses that could fit the facts.

For a thousand years following the eclipse of Theon in A.D. 365, astronomers continued to observe and record solar eclipses. The records become more numerous, and hopefully, more reliable. Newton has found seventeen eclipses between A.D. 840 and 1310 which were recorded by ten or more independent observers. These are listed in Table 1.1. During the 12th and 13th century, 32 eclipses were recorded somewhere in the world, with over 270 records derived from northern Europe alone. When we consider the disturbed state of society and the slow awakening of science during these two centuries, this interest in solar eclipses seems remarkable. However, astronomers did little more than note the location and time of an eclipse, and gave no descriptions of what they actually saw during the event.

Kepler seems to have been the first western astronomer to comment on the appearance of the solar corona, during the eclipse of 1605. The second reference appeared a full century later. Giovanni Cassini, director of the Paris observatory, described a "crown" of pale light surrounding the Sun during the eclipse of 1706. He attributed the light to some manifestation of the zodiacal light.[1]

Nine years later, Edmund Halley observed the eclipse of April 22, 1715, from the roof of the new quarters of the Royal Society of London. In the *Philosophical Transactions of the Royal Society* of that year, he described the shape of the corona and the appearance of the bright red prominences. He noted that they were different on the eastern and western limbs of the Sun, and suggested an explanation in terms of a thin atmosphere on the Moon. According to Halley:

> The eastern limb of the Moon had been exposed to the Sun's rays for a fortnight, and as a consequence it would be natural to expect that the heated lunar atmosphere might exert some absorbing effect on the solar rays, while on the contrary the western edge of the Moon, being in darkness and cold for two weeks, could exhibit no such absorbing action.

Although the idea of a lunar atmosphere was dispelled eventually, it took over 180 years to establish whether the phenomena observed during an eclipse were solar or terrestrial in origin.

DISCOVERY AND DEBATE IN THE NINETEENTH CENTURY

Most of us tend to forget how recently our present knowledge of physics and astronomy has developed. The date 1800, which coincides with the election of Thomas Jefferson as President of the United States, does not seem all that long ago. Yet, at that time, Sir William Herschel, one of the towering figures of eighteenth-century astronomy, seriously proposed the idea that the Sun was a cool, solid body, covered by a layer of luminous clouds, and not greatly different from the planets except for size and distance. It would be 40 years before the law of conservation of energy was proposed, and 70 years before physicists understood how the amount of energy radiated by a hot body varied with its temperature.

Herschel's ideas were rapidly dispelled by scientific investigation during a series of eclipses beginning in 1842. Until that time, scientists had shown little interest in the phenomena visible during an eclipse.

Francis Bailey, an English amateur astronomer, had witnessed the annular eclipse of 1836. He published a report describing a row of bright beads of sunlight that appeared along the Moon's limb at the maximum phase of the eclipse. "Bailey's Beads," as these are now called, are caused by sunlight shining between the mountains on the Moon. His description focused the attention of professional astronomers on the eclipse of 1842, which was visible across southern

Europe. Bailey published the following enthusiastic description of what he saw then:

> I was astounded by a tremendous burst of applause from the streets below, and at the same moment was electrified at the sight of one of the most brilliant and splendid phenomena that can be imagined, for that instant the dark body of the Moon was suddenly surrounded with a *corona,* a kind of bright *glory.* I had anticipated a luminous circle around the Moon during the time of the total obscurity, but I did not expect from any of the accounts of previous eclipses that I had read to witness so magnificent an exhibition as that which took place. Splendid and astonishing, however, this remarkable phenomenon really was, and though it could not fail to call forth the admiration and applause of every beholder, yet I must confess there was at the same time something in its singular and wonderful appearance that was appalling. But the most remarkable circumstance attending the phenomenon was the appearance of three large, *protuberances* apparently emanating from the circumference of the Moon but evidently forming a portion of the corona.

Fifty years of intensive investigation of total solar eclipses followed Bailey's announcement. Before we describe some of the important discoveries made during this period, we need to review a little of the parallel development of the new science of spectroscopy.

SPECTROSCOPY AND SPECTRAL ANALYSIS

Isaac Newton, in 1664, was the first to spread out the colors contained in the white light of the Sun into a "spectrum." He used a simple glass prism. One hundred and fifty years later, Joseph Fraunhofer placed a slit-shaped aperture in front of the prism and so invented the modern spectroscope (see Figure 1.1). When he viewed sunlight through this device, he saw hundreds of dark "lines" scattered throughout the spectrum, as in Figure 1.2. He was the first to measure their position on a numerical scale.

His pioneering work was followed by that of many spectroscopists, including Brewster and Foucault. In the 1860s, Kirkhoff and Bunsen

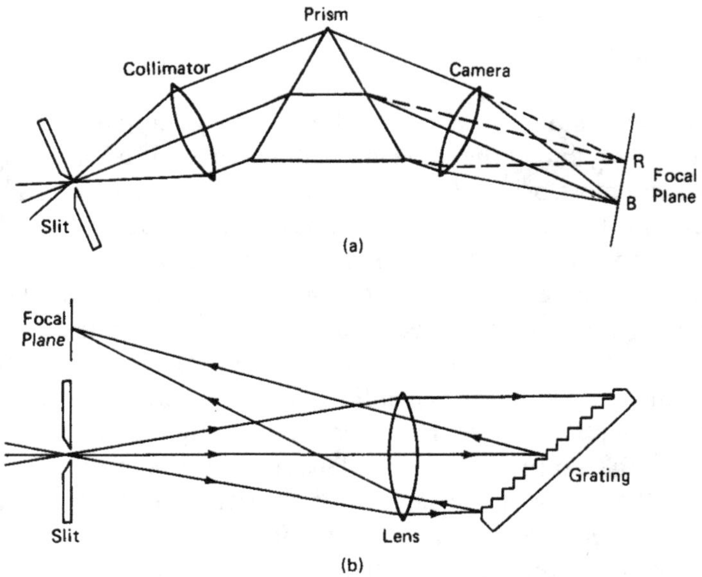

Fig. 1.1. (a) In a simple *prism* spectrograph, light from the source enters through the slit, is spread out into its component colors (or wavelengths) and refocused onto a final photographic plate or film. Blue light (B) and red light (R) are separated, as shown.

(b) In a *grating* spectrograph, the separation of the light into colors is accomplished with a grating. This is an aluminum-coated glass plate, on which fine scratches or grooves have been ruled with a diamond. As many as 1200 grooves/mm can be ruled. The more grooves per millimeter, the wider the separation of the colors.

Fig. 1.2. A section of the spectrum of sunlight, in the yellow. The wavelength of light increases from left to right. The dark lines, discovered by Fraunhofer, are formed in the Sun's atmosphere by atoms of iron and nickel. The bright light portions between the lines is called the "continuum."

mapped the solar spectrum and compared the patterns of dark lines with those produced by terrestrial elements in the laboratory. In this way they were able to identify a score of terrestrial elements in the Sun.

Ångström, in 1869, mapped the solar spectrum with a diffraction grating spectroscope, which produces a linear scale of wavelength. The current unit of wavelength, 1/100,000,000 of a centimeter, is named, appropriately, the Ångström.

Rowland extended Ångström's work. In the 1890s, using concave gratings, he measured the intensity and position of over 20,000 lines in the solar spectrum.

Kirkhoff was the first to propose a convincing explanation for the origin of the lines. As a result of his laboratory experiments, he formulated three rules governing the emission of radiation by solids, liquids, and gasses. First, a heated solid or liquid emits a continuous spectrum, devoid of absorption lines. Second, a hot gas or vapor, composed of a single element, emits bright emission lines at definite, invariable wavelengths. The same gas, when it is cool, will *absorb* radiation at the same wavelengths it emits, i.e. it "reverses" its characteristic spectrum from bright to dark lines.

Kirchoff applied these three physical laws to the Sun. He proposed that the visible surface of the Sun, the photosphere, consists of a hot liquid which emits a continuous spectrum. (Later, prompted by P. A. Secchi, he replaced the liquid by clouds of liquid or solid particles). A cool, gaseous layer overlies the photosphere and absorbs from the continuous spectrum the light that would have emerged in the dark lines. This "reversing layer" produces the Fraunhofer spectrum. Kirchoff's ideas guided experimenters throughout the middle and end of the 19th century.

IS THE CORONA REAL?

For fifty years following the eclipse of 1842, scientists were engaged in a circling debate on whether the phenomena observed during an eclipse (especially the corona) were really part of the Sun or were some kind of optical illusion. Most scientists of the time probably favored an illusion, and ascribed the corona to the scattering of light around the solar limb, or around the lunar limb, or in the Earth's atmosphere.

At the Spanish eclipse of 1860, W. de la Rue and P.A. Secchi used the new daggueréotype plates to photograph the corona from sites 250 miles apart. They compared their photographs after the eclipse and found that many of the features that they had seen appeared in both photographs. They concluded that scattering in the Earth's atmosphere alone could not create the corona, but their evidence was not considered convincing by most other scientists.

At the eclipse of 1868 in India, W. W. Campbell of the Lick Observatory discovered that the pearly white light of the inner corona is strongly polarized. Now, as Figure 1.3 illustrates, white light polarizes when it scatters off small particles through a large angle (say 90°) from its original direction. But light hardly polarizes at all if it scatters into its original direction of travel. Therefore, Campbell concluded that the corona consists of sunlight scattered by small solid particles in the vicinity of the Sun, and not in the Earth's atmosphere. This was a strong argument for the reality of the corona, but Campbell's results were immediately challenged. If his explanation were correct, one would expect to see Fraunhofer lines in the scattered spectrum, and they were not visible. Moreover, in the very next year, 1869, the experiment was repeated, and this time the white light corona was found to be unpolarized!

At the eclipse of 1870, two astronomers repeated the experiment of de la Rue and Secchi of ten years before. Brothers, in Sicily, and Willard, in Spain, each photographed the corona. Their photographs

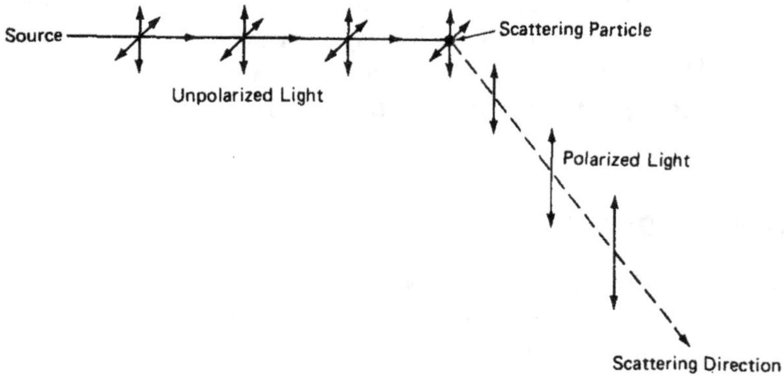

Fig. 1.3. Sunlight is originally unpolarized, i.e. has no preferred direction of vibration. However, when it scatters off a small particle (like a free electron in the corona), it becomes linearly polarized. As shown above, the vibrations are confined to a single preferred line when viewed from any "scattering direction" around the particle.

showed the same details, and reinforced the earlier argument that the corona could not be caused by scattering in the Earth's atmosphere. The possibility that the scattering was produced at the edge of the Moon still could not be ruled out, however.

P. J. C. Janssen contributed to the debate with the observations that he obtained in Hindustan during the eclipse of 1871. He saw Fraunhofer absorption lines, particularly the strong absorption line of sodium in the yellow, superposed on the light of the corona. This evidence supported the idea that the coronal light consists of sunlight scattered by particles near the Sun. But at the same time, Janssen saw emission lines of hydrogen in the inner corona. Kirchoff's laws implied that incandescent clouds of hydrogen must emit these lines. Thus, Janssen's results pointed to *two* components of the solar corona: cool, scattering particles, and hot, emitting gas. While complicating the picture, Janssen's results seemed to favor the reality of the corona.

From 1875 through 1882, the corona was photographed at a series of eclipses, and astronomers noted a systematic change in its shape. In 1871, Janssen had compared the coronal structures to the petals of a giant dahlia, while in 1878, the corona was confined to the equatorial belt of the Sun and extended more than 12 solar diameters along the ecliptic. Now, 1871 corresponded to a maximum in the sunspot cycle (which had been announced by the German amateur, H. Schwabe in 1843), while 1878 corresponded to the dead minimum.

Astronomers realized, therefore, that the corona changes its shape and phase with the sunspot cycle. This was more evidence that the corona is part of the Sun. Moreover, since the corona blends into the zodiacal light, at least its outer portions must consist of small, scattering particles. Janssen's model of a two-component corona has persisted, in modified form, to the present day.

By 1890, the majority of astronomers was convinced that the corona was a real extension of the Sun. The problem then, was to sort out its properties.

THE "CORONIUM" PUZZLE

As mentioned earlier, Kirchoff identified more than a score of terrestrial elements in the Sun by matching laboratory spectra of these elements with the Fraunhofer spectrum. This method rests on the fact that every element, in its gaseous or vapor form, emits a characteristic pattern of spectrum lines at invariable wavelengths. Lines of given elements vary in strength according to definite atomic rules. When the wavelengths and relative strengths of a set of Fraunhofer lines coincide with that of a laboratory sample, we can be sure that element is present in the Sun. Kirchoff's method has developed into one of the most important of astrophysical tools. In the hands of later investigators, the relative *amounts* of different elements, as well as their *presence,* could be determined. In this way, a quantitative chemical analysis of the Sun was developed.

A particular element, such as iron, can emit different characteristic spectra, depending upon the way in which the sample is excited. In a hot furnace, iron will produce one kind of atomic spectrum, but if an iron vapor is subjected to an electrical spark, a more vigorous excitation of the atoms occurs, and a different spectrum appears. We now know that in the spark, iron atoms are ionized; that is, they lose an electron and become positively-charged ions. These ions emit a spectrum that is completely different from that of the neutral atom. Thus, if we did not know anything about ionization, we might conclude that a new element, and not a different form of an element, was present in the spark. Unfortunately, this kind of misunderstanding prevailed throughout the latter half of the 19th century.

At the eclipse of 1869, which was visible over North America, W. Harkness and C. A. Young independently discovered a bright emission line in the spectrum of the corona. Young pointed out that it coincided in wavelength to an iron line in the spectrum of the uneclipsed Sun. He suggested that an iron vapor in the corona must emit the newly discovered emission line.

Astronomers at the time were sure that the coronal atmosphere, if it were real, must be stratified in height according to the weight of its constituents. Thus, they expected to find heavy iron atoms very low in the corona, not at the great height at which Young and Harkness found the new emission line. Later, Young suggested that the new line might be the same as one discovered in terrestrial aurorae. Later still, Young was able to resolve the Fraunhofer iron line in the spectrum of the solar disk into *two* lines, one due to iron and the other presumably due to an unknown coronal element. As it turns out, these wavelength measurements were *also* inaccurate and Young badly misled the subsequent course of the investigation.

The number of coronal emission lines without satisfactory identifications continued to accumulate. At the eclipses of 1882 in Egypt, of 1883 on Caroline Island in the Pacific, and of 1886 in the West Indies, A. Schuster and W. de W. Abney photographed about 30 bright lines, of which at least two in the violet belonged to the corona. A. Fowler photographed nine coronal lines at the West African eclipse of 1893 whose wavelengths agreed with those reported for the eclipse of 1886. Also in 1893, Des Landres photographed six coronal lines.

At this point, ten coronal lines were known and, with the possible exception of Young's coronal line, no identification with a terrestrial element was possible. A new element, "coronium," was postulated to account for them. After the Indian eclipse of 1898, even Young's suggested identification with iron fell apart. Fowler obtained coronal spectrograms of high precision during the eclipse. When J. N. Lockyer measured them, he found a wavelength of 5303 A, in our modern notation, for the "coronium" line, whereas the Fraunhofer line that Young had associated with the coronium line lies at 5317 A, 14 angstroms away. Young had caused a great deal of confusion simply because his equipment was not sufficiently precise for the task.

Five new coronal lines were discovered at the eclipse of 1914, including the second strongest in the entire spectrum, at 6374 A. By 1927, 16 coronal spectrum lines ranging in wavelengths from 3388 A to 6374 A, were known. No identification for any of these lines was possible. Astronomers had to wait for another 15 years before the puzzle of "coronium" was solved, and the true nature of the corona began to emerge.

PROMINENCES AND THE CHROMOSPHERE

Astronomers had far less difficulty in accepting prominences as a real extension of the solar atmosphere than they did the corona. Part of the reason may be that prominences *look* like bright red clouds that extend beyond the Moon's limb at the moment of totality (see Figure 1.4). Once the idea of a lunar atmosphere was discredited, the way was open to accept "protuberances" or prominences as extensions of the Sun's atmosphere. The idea seemed more plausible after the Scandanavian eclipse of 1851 when the Moon was observed to cover and uncover prominences in its motion across the Sun.

The first real understanding of the nature of prominences followed

Fig. 1.4. The total eclipse of February 16, 1980, photographed by a team from the High Altitude Observatory. The bright areas just on the dark limb of the Moon are solar prominences. (*Courtesy of the High Altitude Observatory, National Center for Atmospheric Research, sponsored by the National Science Foundation, and of Southwestern at Memphis.*)

the Indian eclipse of 1868. The eclipse attracted large expeditions from Austria, Britain, France, and Germany. The French astronomer Janssen viewed the spectrum of prominences during the eclipse, and was struck by the appearance of half a dozen bright emission lines. The brightest lines coincided in wavelength with Fraunhofer lines which had already been identified with hydrogen. So Janssen concluded that prominences must be predominately comprised of hot hydrogen gas. Janssen also noted a bright emission line in the yellow, which most other observers identified with the well-known sodium "D" line, but which Janssen realized was at a slightly shorter wavelength.

Janssen was impressed with how bright the prominences appeared during the eclipse, and he conceived the idea of searching for them the day after. He set the slit of his spectroscope at the limb of an image of the Sun, and to his great surprise and pleasure, saw the bright hydrogen emission line at the same position where a prominence had appeared during the eclipse. He dispatched a letter to the French Academy of Science, in which he wrote:

> Prominences of the day before were considerably reduced; there was hardly any sign of the large prominence and the distribution of the gaseous matter was quite different. From that day until 4 October I have constantly studied the Sun from this point of view. I have drawn maps of the prominences which show how quickly (often in minutes) these huge gaseous masses change shape and position. Finally during this period, which was like a seventeen-day eclipse, I collected a considerable number of facts which in themselves indicate the physical composition of the Sun.

J. Norman Lockyer, the eminent British astronomer, had been unable to attend the eclipse; he was awaiting the delivery of a new and more powerful spectroscope, which arrived in October of 1868 at Cambridge University. Lockyer had guessed, independently of Janssen, that prominences might be visible through the spectroscope without an eclipse. When his new instrument arrived, he trained it onto the Sun, in broad daylight, and immediately confirmed his suspicion. He wrote straight away to the French Academy of Sciences. By an incredible coincidence, his letter arrived within a few minutes of Janssen's, and they were read, one after the other, at the same meeting.

On November 15, 1868, Lockyer definitely established that the yellow prominence line differed in wavelength from that of sodium. He collaborated with Edwin Franklin, a chemist, in a long series of experiments to find a terrestrial element that would emit a spectral line at a wavelength of 5876 A. After many failures, Lockyer was convinced that he had discovered a new element, present only in the Sun and not on the Earth, and called it Helium, after the Greek work "helios" for Sun. Even his partner, Franklin, was not quite convinced, however. He was left with the possibility that hydrogen, at some extreme solar temperature or pressure, might emit the new line.

Astronomers were understandably reluctant to accept a new element on the basis of a single spectral line. It was not until 1895, when W. Ramsay reproduced the solar line in the laboratory from a sample of gas obtained from the mineral cleavite, that helium became firmly established as a new element.

The study of the solar chromosphere began at the eclipse of 1851, when G.B. Airy described it as a ring of bright red emission, with a rough outer edge. Airy named this region the "sierra," thinking that it might represent mountains on the Sun. Herschell's concept of a solid solar surface dissipated during the 1860s, and eventually "sierra" was replaced by "chromosphere."

After the eclipse of 1868, Lockyer and Secchi studied the chromosphere spectroscopically outside of the eclipse and measured the heights to which the strongest emission lines reached, but it was left to C. A. Young to connect these emission lines to Kirchoff's idea of a "reversing layer." Kirchoff's rules state that a hot gas will absorb and emit light at the same wavelengths. The emission becomes apparent when the gas is projected against a dark background, like the black sky at an eclipse.

During the Spanish eclipse of 1870, Young saw the photospheric Fraunhofer lines change over into emission. He described his experience as follows:

> As the Moon advances, making narrower and narrower the remaining sickle of the solar disk, the dark lines of the spectrum for the most part remain sensibly unchanged, though becoming somewhat more intense. A few, however, begin to fade out, and some even begin to turn palely bright a minute or two before totality begins. But the moment the Sun is hidden, through the whole length of the spectrum—in the red, the green, the violet—the bright lines flash out by hundreds and thousands almost startlingly; as suddenly as stars from a bursting rocket head, and as evanescent, for the whole thing is over in two or three seconds. The layer seems to be only something under a thousand miles of thickness, and the Moon's motion covers it very quickly.

After this eclipse, Young, following the lead of Janssen and Lockyer, mapped the chromospheric emission spectrum out of eclipse. He recorded over 300 different lines and noticed that many of them appeared sporadically in the slit of his spectroscope, as though material was ejected upward from the solar surface. Moreover, although there was a generally good correspondence between dark Fraunhofer lines and bright chromospheric lines, some were missing and some were strongly enhanced. The reasons for all these phenomena only began to emerge in the 1930s.

The study of the flash spectrum was held up until photography developed sufficiently. The first photograph was secured in 1883 and was followed throughout the 1890s by continuously improving records.

W. W. Campbell, at Lick Observatory, invented a new technique for photographing the flash spectrum and, in particular, the height to which different lines reach. He mounted a long slot in front of the photographic plate in the direction of wavelength. During the eclipse, he arranged the entrance slit of the spectrograph along a radius of the Sun. In this way, the full height of the chromosphere at any moment is imaged through the slot onto the photographic plate. Then as the Moon advanced, slowly covering up the bright chromosphere, the photographic plate was moved under the slot.

A typical spectrogram is shown in Figure 1.5. The vertical direction (perpendicular to the wavelength scale) corresponds to *time* during the eclipse or, equivalently, height. Such a record displays very compactly the behavior of each line near the limb as it crosses the transition from absorption to emission.

Campbell applied this method to the eclipses of 1898, 1900, and 1905. D. H. Menzel analyzed these records and was the first to extract from them useful astrophysical information.

THE CORONA EXPLAINED

The most important problem facing solar astronomers at the beginning of the 20th century was the identification of the element coronium, and the explanation of the conflicting evidence on the temperature and structure of the corona. You can get some idea of how confused solar astronomers were from a discourse given by Professor Arthur Schuster, at the Royal Institution in London in 1891. Schuster listed four alternatives for the nature of the corona:

> It consists of matter either (1) forming a regular atmosphere around the Sun, or (2) matter projected from the Sun, or (3) matter

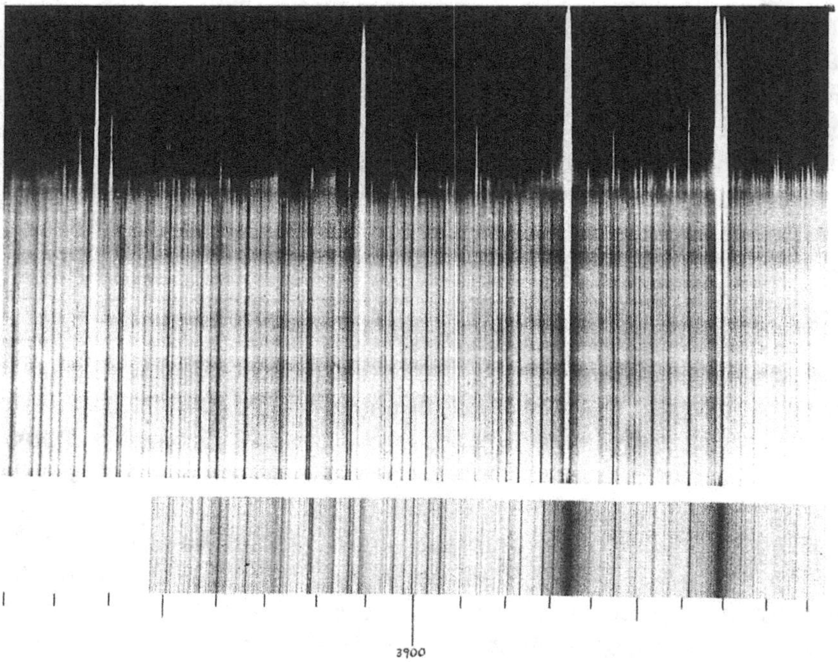

Fig. 1.5. The upper panel shows a section of the chromospheric flash spectrum, photographed with Campbell's "moving plate" spectrograph during the eclipse of 1905. Wavelength increases from left to right, and the radial direction on the Sun is vertical. Inside the limb (bottom of panel), the Fraunhofer lines are in *absorption,* i.e. dark. At the limb and outside it (top of panel) the lines go into *emission* against the black sky. A photospheric spectrum appears in the lower panel.(*Lick Observatory photograph.*)

> falling into the Sun, or finally (4) matter circulating around the Sun with planetary velocity. We may at once reject the first and fourth, for it may be proved that the Sun could have no regular atmosphere to the extent indicated by the outlines of the corona, and spectroscopic results exclude the hypothesis that the bulk of its matter revolves with planetary velocity, though probably there is some meteoric material which does revolve around the Sun.

The problem of the corona was solved over the next 40 years with assistance from the fields of stellar astronomy and atomic physics.

The spectra of gaseous nebulae contain a number of emission lines not emitted by any terrestrial material, under laboratory conditions. A hypothetical element, "nebulium," was postulated to account for them. By the mid-1920s, however, the Periodic Table of the elements was pretty well filled up and had no room for such a mysterious element, any more than it did for "coronium."

The riddle of nebulium was solved by I. S. Bowen, a physicist at the California Institute of Technology. He showed theoretically that an extremely dilute gas emits lines which are ordinarily forbidden, or at least highly improbable, under the laws of quantum physics. He identified some of the strongest lines of the nebular spectrum as forbidden lines of singly-ionized and doubly-ionized oxygen and singly-ionized nitrogen. In 1935, he identified some puzzling lines in the spectrum of Nova Pictoris 1925 as forbidden lines emitted by iron atoms that were missing 5 or 6 outer electrons.

You will remember that the spectra emitted by a neutral atom and by the ions it forms when it loses one or more electrons are completely different. By the early 1930s, spectroscopists understood the energy levels of bound atomic electrons sufficiently well to predict, with reasonable accuracy, the permitted and forbidden emission lines that such ions would emit. This was the basic advance in physics that enabled Bowen to score his successes.

In 1933, a recurrent nova in the constellation Ophiuchius exploded. Adams and Joy, at the Mt. Wilson Observatory, identified five solar coronal lines in its spectrum. In Germany, W. Grotrian put two and two together, and suggested that the coronal lines in the nova might also be emitted by forbidden transitions of highly ionized elements. The Swedish physicist B. Edlen had been studying the energy levels of highly ionized atoms, including nine-times-ionized iron (Fe X). In 1939, Grotrian used Edlen's results to demonstrate that the red line emitted by the solar corona at a wavelength of 6374 A could be attributed to Fe X.

This was the breakthrough that was needed. Edlen picked up Grotrian's suggestions, and by 1942 had identified 19 of the 27 known coronal emission lines as forbidden transitions in highly-ionized atoms of such common elements as calcium, iron, and nickel.

Now the situation was becoming clear. In 1920, M. Saha had developed the theory of ionization of atoms in a dilute gas. He showed that the number of electrons stripped off from a complex neutral atom increases with the temperature of the gas. At any particular temperature, an element such as iron exists concurrently, in the form of neutral atoms, ions with one electron missing, ions with two electrons missing, and so on. The distribution of the element among its possible ionic states changes in step with the gas temperature (see Figure 1.6). Conversely, the most abundant ion of a given element is an indicator of its temperature. Iron atoms with nine electrons missing (Fe X) can exist in abundance only at a gas temperature of a million degrees Kelvin, for example.

Thus, the work of Grotrian and Edlen led directly to the conclusion that the solar corona is a dilute gas with an extremely high temperature. This immediately accounts for its enormous extension, which so puzzled the astronomers of the 19th century. It also accounts for the absence of Fraunhofer lines in the white light of the corona. This "continuum," as it is called, consists of sunlight scattered from extremely hot coronal electrons, whose motion blurs out any spectral detail present in the incident sunlight.

In order to account for the presence of Fraunhofer lines in the light of the *outer* corona, one has to invoke the presence of cool, small particles. Meteoric particles orbiting the Sun, in the plane of the ecliptic, satisfy this requirement quite well, and provide a common explana-

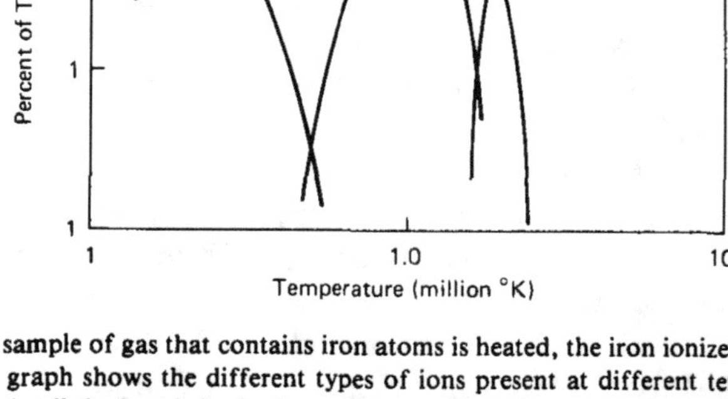

Fig. 1.6. As a sample of gas that contains iron atoms is heated, the iron ionizes, losing bound electrons. The graph shows the different types of ions present at different temperatures. At 600,000°K nearly all the iron is in the form of ions with 7 electrons missing (Fe VIII). As the temperature rises, ions of Fe X, XI, etc. predominate. *(from Carole Jordan, "The Ionization of Equilibrium of Elements between Carbon and Nickel," in* Monthly Notices of the Royal Astronomical Society, *Vol. 42, 1969, by permission of the Royal Astronomical Society.)*

tion for the dust component of the corona and for the zodiacal light (see Chapter 6).

In summary, the corona was now viewed as a hot, tenuous gas consisting mainly of ionized hydrogen in the form of electrons and protons, with minute traces of heavier elements such as calcium and iron. This large but very thin corona is imbedded in a cloud of cool solid particles. Later studies of the white light of the corona have established that the density of electrons and protons is far lower by many factors of ten than the best laboratory vacuum.

THE STRANGE PHYSICS OF THE CHROMOSPHERE

A series of investigations culminating in 1942 established that the chromosphere, like the corona, was much hotter than anyone had imagined, and behaved unlike any gas known in the laboratory.

D. H. Menzel picked up the analysis of W. Campbell's moving plate photographs of the flash spectrum, taken at the Spanish eclipse of 1905. Using an ingenious technique, Menzel determined the density of radiating atoms and ions of several important elements in the chromosphere from the intensity of the spectral lines that they emit. According to L. Boltzmann, the German physicist, the number of atoms or ions excited to each of its possible energy states (see Figure 1.7) should follow an exponential law:

$$N = N_0 e^{-E/kT}$$

where T is the temperature of the gas, E is the energy of the excited electron within the atom, and k is a universal constant, named after Boltzmann.

When Menzel applied this formula to the eclipse data, he found that mildly excited atoms (with values of E smaller than about 2 electron volts), yielded a temperature of about 4000°K. On the other hand, highly excited atoms yielded a temperature of 13,000°K. The same gas, of course, cannot have two temperatures if it is in thermodynamic equilibrium.[2] Menzel concluded that the chromosphere could not be in thermodynamic equilibrium.

Next, Menzel determined the rate at which the density decreases with increasing height, or the "scale height." The scale height is the

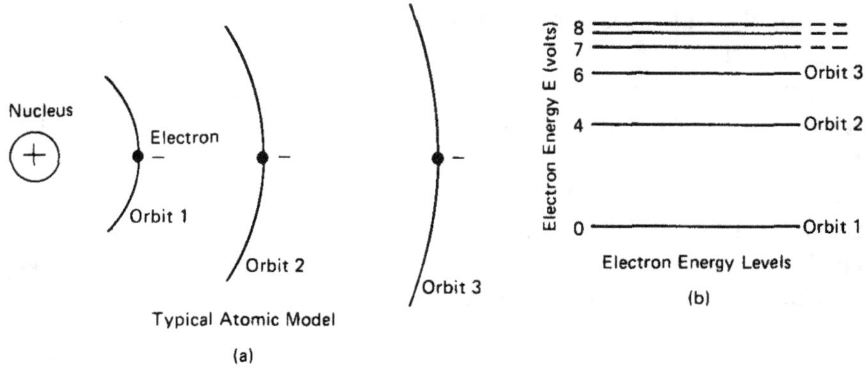

Fig. 1.7. (a) Electrons in a complex atom occupy discrete orbits around a central positive nucleus. An electron has a definite potential (electrostatic) energy in each particular orbit.

(b) An electron in orbit 1 has the lowest energy, in orbit 2 a slightly higher energy, and so on. If an electron jumps from one orbit to a lower orbit, the atom emits a photon of light, whose energy is the difference in the energies of the orbits.

distance in which the density falls by a factor of 2.7. This quantity also depends on the temperature, as you might expect, with large scale heights corresponding to high temperatures. Menzel found a scale height of 400 km at a height in the chromosphere of 2000 km, which corresponds to a temperature of 6300°K. Here was more evidence that the temperature of the chromosphere was higher than that of the photosphere.

This was an idea that astronomers were unwilling to accept. To account for the facts, they imagined several special processes that might somehow support the chromosphere against gravity.

For example, W. H. McCrea proposed "turbulent pressure." If the gas were churning sufficiently vigorously, the motion would generate a pressure of its own in addition to the static gas pressure, and this extra pressure would elevate the chromosphere. Turbulent motions as large as 18 km/sec were necessary to account for the observations, but the sound speed at 6000° is only 12 km/s, so the motions would be supersonic, would tend to dissipate in shocks, and would heat the chromosphere.

Before 1930, solar astronomers were content to obtain one or two photographic spectra of the chromosphere during an eclipse. They would lay the slit of their spectrographs either tangent or radial to the

Moon's limb, snap off a picture, and hope the exposure was right. They would never quite know at what height the slit was located. Campbell's moving plate spectrograph was an improvement in this regard, because it recorded the entire spectrum throughout the flash phase, but only for one point on the limb of the Sun.

Menzel and J. F. Chappel invented a new kind of spectrographic camera, the "jumping film" camera, for the Canadian eclipse of August 31, 1932. No slit was used with this spectrograph; instead, the thin crescent of chromosphere that appears luminous during the flash phase is used as a slit-like object itself. Between successive exposures, the camera advances the film very rapidly. In this way, as the Moon covers or uncovers the chromosphere, a rapid series of spectra covering a wide arc at the limb of the Sun is recorded. The wavelength resolution of such a spectrograph is too small to resolve the intrinsic widths of the lines. As Figure 4.3 (see page 61) shows, the flash spectrum consists of many bright arcs, one for each spectral line. The line width is determined partly by the instrument and partly by the intrinsic height of the chromosphere.

G. G. Cillie and Menzel analyzed the 1932 eclipse spectra and showed how modern astrophysical techniques could extract a wealth of physical information from such records. For example, they demonstrated that the gas pressure falls off by a factor of 400 between the top of the photosphere and the base of the chromosphere, and that the gas pressure influences the appearance of the spectrum as well as the gas temperature. This result agreed with the famous Saha formula for the ionization equilibrium of a given element.

Cillie and Menzel found that a gas temperature of about 10,000°C was sufficient to explain their observations of the strengths of spectrum lines, once they determined the gas pressure. The chromosphere was definitely hotter than the photosphere.

The big surprise came after the 1940 eclipse in South Africa. The Cambridge University astronomer, R. Redman, obtained a superb set of flash spectrograms using a slit spectrograph. When he measured the *widths* of metal line profiles, he found they implied a chromospheric temperature of no less than 30,000°K, three times the temperature found by Cillie and Menzel by a different method! Turbulent motions in the chromosphere were certainly no larger than 1 km/sec, 20 times smaller than McCrea had proposed.

Astrophysicists were now faced with a puzzle: the chromosphere presented a different "temperature" depending upon how you measured it. Spectrum line *intensities* implied *low* excitation and ionization temperatures (5000 to 10,000°K). Line *widths*, on the other hand, which indicate the motions of the gas, and therefore its true kinetic temperature, indicated a much higher temperature.

This region of the Sun would present severe challenges to theoretical astrophysicists, who would struggle all through the 1950s and 1960s to develop a theory of an opaque gas that disobeys the laws of thermodynamic equilibrium.

We have retraced 30 centuries of eclipse science, at breakneck speed, and reached the modern era of research. Before continuing with recent experiments, we will take a moment, in the next chapter, to explain the beautiful celestial machinery that makes eclipses possible. Then we will resume our story of the chromosphere and corona, interplanetary medium, and the atmosphere of the Earth.

NOTES

1. Zodiacal light is visible as a long cone of faint light that extends along the ecliptic in the western sky after sunset. It consists of sunlight scattered by minute meteoric particles that are concentrated in the plane of the Earth's orbit.

2. A system (like a sample of hot gas) reaches "thermodynamic equilibrium" when all chemical, mechanical, and energy transport processes stabilize so that there is no further tendency for spontaneous change. If the gas were enclosed in rigid, nonconducting, perfectly reflecting walls, for example, it would tend toward this condition. All microscopic processes (such as atomic collisions or creation of photons), and the spectrum of radiation within the gas, would then be characterized by a single temperature, and universal laws would govern the gas properties (e.g. ionization).

   In the Sun's atmosphere this condition is approached only in deep (subphotospheric) layers, that are effectively isolated from outer space. However, the outer layers (e.g. chromosphere and corona) never reach this state. Their properties are determined by the rates of individual microscopic processes (collisions, etc.) and these may *appear* to run at different temperatures.

SUGGESTIONS FOR FURTHER READING

Books

Lovell, B. (ed.) *Library of Science, Vol. I.* Elsevier Publishing Company: 1970.
Meadows, A. J. *Early Solar Physics.* Pergammon Press: 1970.
Mitchell, S. A. *Eclipses of the Sun.* Columbia University Press: 1951.
Needham, J., and Ling, W. *Science and Civilization in China.* Cambridge University Press: 1959.
Newton, R. R. *Ancient Astronomical Observations.* Johns Hopkins University Press: 1970.
Pannekoek, A. *History of Astronomy.* Interscience: 1961.
Stephenson, F. R., Clark, D. H. *Applications of Early Astronomical Records.* Oxford University Press: 1978.

Periodicals

Chou, H. "Oracle Bones," in *Scientific American,* April 1979.
Newton, R. R. *Science* **166,** 825, 1969.

2
Eclipses Explained

Nearly everyone has a rough idea of why solar eclipses occur. If asked, the average man in the street might reply, "Because the Moon moves between the Sun and the Earth." And he would be right, except that his answer glosses over the incredible accident that the Moon and the Sun appear to have the same size, even though the Sun is actually 400 times larger than the Moon. It is the nearly exact equality of the angular diameters of the Sun and Moon, as seen from the Earth, that makes total eclipses possible at all.

THE MOON'S SHADOW

As Figure 2.1 shows, the Moon's shadow has two parts: (1) the *umbra,* in which no sunlight falls; and (2) the *penumbra,* in which light from only some parts of the Sun falls. The length of the Moon's umbra varies only slightly, from 228,000 to 236,000 miles. On the other hand, the Moon's distance from the Earth's surface varies appreciably (217,000 to 249,000 miles) because the Moon's orbit is fairly elliptical.[1] If you compare these numbers, you will see that the tip of the umbra may sometimes reach or extend beyond the Earth's surface. When this happens, the umbra touches the surface in an elliptical spot, labeled TT′ in Figure 2.1. Any observer within the spot will see a *total eclipse* of the Sun. When the Sun, Moon, and Earth line up precisely, the spot contracts to a circle, and when the Moon is at perigee (nearest the Earth) during the eclipse, the maximum diameter of the circle is about 167 miles.

If the tip of the Moon's umbra falls above the Earth's surface (when the Earth is at E_2 in Figure 2.1) a total eclipse is not possible, but

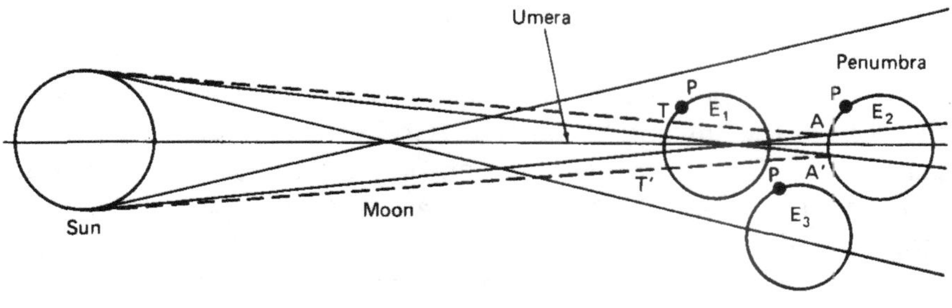

Fig. 2.1. Geometry of the Moon's shadow for total, annular, and partial eclipses (see text for explanation).

anyone standing in the *extension* of the umbra—that is, in the spot marked A–A′ in Figure 2.1—will see the Moon projected against the Sun as a smaller body, and will enjoy an *annular* eclipse. An eclipse may be total along part of the eclipse path and annular along another part.

Finally, if the Earth lies in the penumbra, but not in the umbra (as at E_3 in Figure 2.1), you will see a *partial* eclipse if you watch from a typical point "P". Moreover, if you happen to stand outside the path of an annular or total eclipse, as at the points P on E_1 and E_2, you will still see the Moon covering part of the sun, i.e., a partial eclipse.

A total eclipse progresses as shown in Figure 2.2. In the Northern Hemisphere, the eclipse usually occurs in the southern half of the sky, with east on the left and west on the right. Both the Sun and Moon appear to move westward because of the rotation of the Earth. At the same time, because of the Earth's orbital motion, the Sun appears to crawl eastward (1° per day), relative to the distant stars. However, the Moon moves eastward about 13 times faster than the Sun, because of

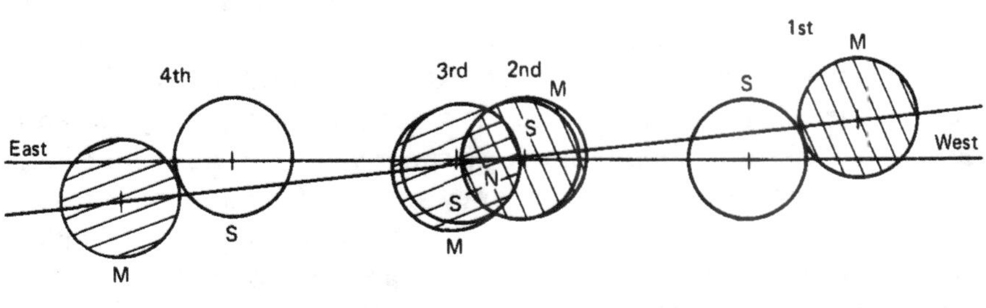

Fig. 2.2. The "contacts" of a total eclipse occur in sequence as the Moon overtakes the Sun. The sequence begins with first contact on the right and ends with fourth contact on the left.

its rapid orbital motion.[2] So, as we watch, we see the Moon overtake the Sun, coming from the west, gradually covering the Sun and moving off to the east. Figure 2.2 shows the various "contacts" between the disks of the Sun and Moon. The last sliver of the solar disk disappears just before second contact, and the total phase of the eclipse lasts until third contact. Only during this time can we see the outer atmosphere of the Sun, nearby stars, planets, or comets.

The length of totality clearly depends on the relative apparent sizes of the Sun and Moon and their relative motion. Their sizes vary from eclipse to eclipse because of the ellipticity of their orbits. Their motions vary from one eclipse to another because the Moon moves faster when it is near perigee (the point in its orbit nearest the Earth) than near apogee (the point furthest from the Earth).[3] The maximum ratio of the Moon's angular diameter to the Sun's angular diameter defines the *magnitude* of the eclipse. The magnitude is larger than 1.0 for total eclipses and less than 1.0 for annular or partial eclipses.

It is straightforward but tedious to describe how to calculate the maximum difference in the apparent diameters of the Sun and Moon and the maximum duration of totality. The maximum diameter difference is 2′38″ and the maximum duration of totality is 7 minutes and 40 seconds for an observer near the equator. The 1973 eclipse in West Africa came very close to this maximum theoretical totality. On the average, a total eclipse only lasts for two or three minutes and *seems* much shorter. However, the total duration of an eclipse, from first through fourth contacts, may last as long as four hours.

The umbra of the moon sweeps along the Earth from west to east in a long, narrow path. If it were not for the rotation of the Earth, the leading edge of the shadow spot would pass an eclipse site at more than 2100 miles an hour, but since the Earth rotates in the same direction as the shadow, the relative speed of the spot is smaller. At the equator, the Earth rotates at 1,040 miles per hour, so the shadow's speed drops to 1,060 miles per hour. At higher latitudes, the Earth's rotation speed is smaller, so the shadow's relative speed is higher, and the relative speed of oblique shadows can reach 5,000 miles an hour.

Total eclipses are not only short, they are rare. Let us see one reason why. The area of a typical eclipse path, 6,000 miles long and 100 miles wide, is 600,000 square miles. The Earth's area, by comparison, is 197 million square miles. Thus if you insist on living at one place on the Earth and not traveling at all, the probability that a total eclipse will pass over your town is 600,000/190,000,000 or 1/305. As we shall learn later, total eclipses occur on the average about every 1.5 years. With a probability of 1 in 300, you would have to wait *450* years for an eclipse to come to you. No wonder eclipse expeditions are popular!

ORBITS AND PERIODS

The gravitational forces of the Sun and planets perturb the Moon so that it never retraces exactly the same orbit on any two successive passes. The Moon's orbit in three-dimensional space is exceedingly complicated, but it can be described fairly accurately as an ellipse of approximately fixed size and shape that turns slowly in space (see Figure 2.3.) The ellipse inclines to the Earth's orbital plane by an angle that averages 5° 8′ and varies periodically by only about 8′. The major axis of the ellipse rotates in the plane of the Moon's orbit, counterclockwise as seen from north of the Earth's orbital plane, completing a rotation in 9 years. Earth and Moon revolve in their orbits in the same "prograde" direction (west to east) as the Earth spins. The two orbital planes intersect in a line (the line of nodes) which points in *approximately* a fixed direction among the distant stars as the Earth revolves in its orbit. Actually, the line of nodes "regresses" (rotates westward, or clockwise as seen from north of the Earth's orbital plane) and completes a revolution in 18.6 years.

We always see the same face of the Moon because it rotates on its axis in exactly the same time that it completes a revolution around the

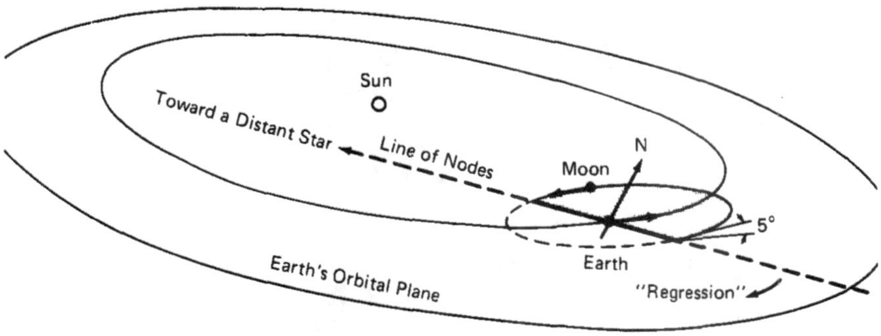

Fig. 2.3. The plane of the Moon's orbit intersects the plane of the Earth's orbit in the "line of nodes," which rotates once in 18.6 years toward the west.

Earth. Because the *rotation* is uniform, whereas the angular speed of revolution is nonuniform (see Note 2), the Moon seems to rock slightly (or librate) both in longitude and latitude, so that we actually see about 60% of the Moon's surface, not simply half.

As the Moon revolves in its orbit, we see it change *phases*. Figure 2.4 should help to remind you that the lunar phases (new, first quarter, full, and last quarter) correspond simply to the fraction of the Moon's disk we see illuminated by the Sun. When new, the Moon lies in the direction of the Sun but it does not *necessarily* lie in the Earth's orbital plane. In Figure 2.4, for example, the solid line portion of the Moon's orbit lies above (north of) the Earth's orbital plane, because the Moon's orbit is tipped 5° with respect to that plane. Thus the new moon lies above the plane. As seen from the Earth, this new moon would appear about 3° north of the Sun, too far to eclipse the Sun. When the Moon has revolved counterclockwise to the "descending" node, where it passes from north to south of the Earth's orbital plane, it lies *in* that plane, but it is no longer new. It is approaching first quarter, in fact.

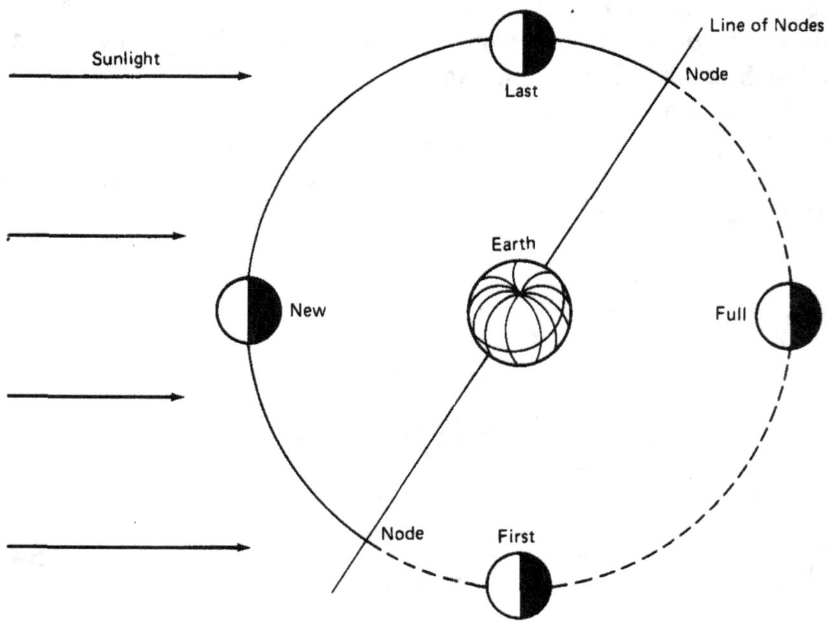

Fig. 2.4. Phases of the Moon.

We see, therefore, why the Moon does not eclipse the Sun each time it becomes new. For a solar eclipse, *two* conditions must be met: the Moon must be new *and* lie near a node.

When are both conditions met? Consider Figure 2.3 again. As the Earth revolves around the Sun, the line of nodes continues to point in a fixed direction among the stars, if we neglect its regression. Therefore the line will point directly from Earth to Sun only twice a year, six months apart. These periods are properly called the "eclipse months," because if a new moon occurs during such a period it will necessarily fall near a node, and an eclipse is likely.

However, one more condition must be satisfied for an eclipse to occur. To understand it, we must shift our point of view. Instead of watching from a point outside the solar system as in Figure 2.3, let us imagine watching the Moon and Sun from Earth as they move against the background of the stars. As we recalled earlier, the orbital motion of the Earth causes the Sun to appear to move eastward with respect to the stars along a track in the sky called the "ecliptic." Similarly the Moon's orbital motion carries it along a track in the sky that we will call the "Moon's path." Figure 2.5 shows the Sun and Moon near the intersection of these two tracks, which we will call the *projected* node. It is the projection on the sky of the *real* intersection, in space, of the lunar orbit with the Earth's orbital plane. The Moon's path intersects the ecliptic at an angle of 5°, the tilt angle of the Moon's orbital plane.

The Sun appears to move with respect to the distant stars about 1° eastward per day, and makes a circuit of the sky from a particular star back to the same star in a sidereal year (365.25 mean solar days). Since the line of nodes regresses westward, the projected node shown in Figure 2.5 is moving westward to meet the Sun as it advances eastward. Thus the Sun takes only 346.62 days (an eclipse year) to travel from one node on the ecliptic back to the same node. Similarly,

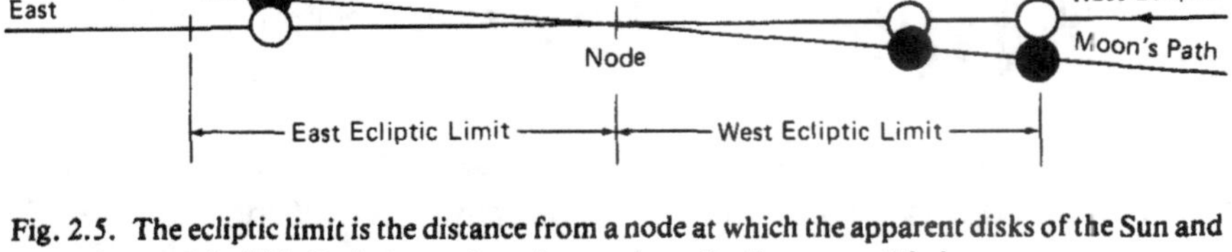

Fig. 2.5. The ecliptic limit is the distance from a node at which the apparent disks of the Sun and Moon just touch. This limit determines the number of eclipses possible in a year.

the Moon's period with respect to a projected node (the "draconic" month, 27.21 days) is slightly shorter than the sidereal month (27.32 days) and about two days shorter than the synodic month (29.53 days), the interval between successive new moons.[4]

In the eclipse months, when the line of nodes points from Earth to Sun, the Sun passes a projected node. Then there is a chance that the Moon, moving 13 times faster than the Sun toward the east, will overtake the Sun at new moon and that an eclipse will occur. The maximum distance of the Sun from a projected node at which an eclipse is just possible is defined as the *ecliptic limit.* (See Figure 2.5.) At this distance, the disks of Sun and Moon just touch. The size of the ecliptic limit obviously depends on the inclination of the Moon's path to the ecliptic and on the apparent angular sizes of the Sun and Moon. These three quantities can be predicted so that a little simple trigonometry will then determine the size of the ecliptic limit. It varies from a "major" value of 18.51° to a "minor" value of 15.35°.

And now finally we come to an interesting conclusion. Since the Sun moves 30.67 degrees relative to a node in a synodic month [29.53d × 360°/346.62d] and since this distance is *less* than twice the minor ecliptic limit (30.70°), we conclude that the Sun cannot pass through the double minor ecliptic limits without being overtaken by a new moon, and that therefore *one eclipse is inevitable at each node.* Thus, the final condition for an eclipse to occur is that the Sun lie within a minor ecliptic limit of a projected node.

If conditions are especially favorable, *two* eclipses are possible at each node, as shown in Figure 2.5. When the apparent sizes of the Sun and Moon are largest and the inclination of the Moon's path to the ecliptic is smallest, the *major* ecliptic limit determines how many eclipses will occur at each node. Imagine that an eclipse has occurred slightly east of the western major ecliptic limit. The Moon will be new again in a synodic month (29.53 days) and will appear 30.6° east of its last position [(29.53 - 27.21) days × 360°/27.21 days]. Thus the Moon will overtake the Sun before either has passed through the double major ecliptic limits (2 × 18.5° = 37°). A second eclipse must occur. Under these conditions two eclipses will occur at each node for a total of four eclipses per year.

Finally, we may get an extra bonus eclipse within a calendar year if the very first one occurs in January. The reason is that there will be 13 new moons in that year. The next time this happens is in the year 2160, however, so you can relax!.

Summing up, we find that it is possible to have two, three, four or five solar eclipses in a calendar year.

Now we can appreciate why eclipses occur in a *series,* at 6 month intervals. Suppose an eclipse has occurred at the ascending node. After 6 synodic months, the new moon will appear (6 × 30.67° = 184.02°) east of the ascending node or 4.02° east of the descending node. The Sun, meanwhile, has also moved 4.02° east of the descending node, so the *next* eclipse in the series occurs. The series will continue with eclipses every 6 synodic months at alternating nodes. Since each eclipse shifts 4° east and the double major ecliptic limits are 36°, a series will contain 9 eclipses at most. A series begins with 1 or 2 partial eclipses near the western eclipse limit, and ends with 1 or 2 partial eclipses near the eastern limit. In the middle of the series, near the nodes, 4 to 6 central eclipses take place.

One synodic month before the current series ends, a new moon will occur inside the western eclipse limit (i.e. 30.67° west of the last new moon in the series). Thus the *next* series begins, overlapping the present one, and shifted earlier by one synodic month.

The 9 eclipses in a series bear little family resemblance. They are visible at widely separated places on Earth and, of course, at 6 month intervals. In contrast to this "normal" series, successive eclipses in a saros family are very similar.

RECURRENCES AND CYCLES

As we saw in Chapter 1, the saros was known to the ancient Greeks and possibly to the Chaldeans. The saros derives from a remarkable natural coincidence: 223 synodic months (a synodic month is the interval between two new moons) equals 6585.32 mean solar days, while 19 eclipse years (the time it takes the Sun to move from one projected node around the sky back to the same node) equals 6585.78 days, or 18 years 11.3 days. The near equality of these two crucial periods guarantees that the configuration of the Sun and Moon with respect to the projected nodes nearly repeats every 18 years. As a result, eclipses repeat, with only slight changes, every 18 years.

Table 2.1 lists some past and future eclipses that are repetitions of

Table 2.1 Some Past and Future Eclipses at Intervals of a Saros

		MAXIMUM DURATION		TURNING POINT OF PATH	
DATE	TYPE	MIN	SEC	LATITUDE	LONGITUDE
May 20, 1947	Total	5	14	+ 6.14°	+ 1.43°
May 30, 1965	Total	5	16	+ 0.93°	+116.11°
June 11, 1983	Total	5	11	− 4.70°	−126.91°
June 21, 2001	Total	4	56	−10.97°	− 8.73°
July 2, 2019	Total	4	32	−17.48°	+104.94°

each other at intervals of a saros. Notice the following strong regularities in this sequence. First, successive eclipses fall in the same eclipse season, spring, because all eclipses in the series occur at the *same* (ascending) node. Of course, each eclipse date advances 11 days, so the eclipse season dates gradually shift. Second, each eclipse path falls about 120° *west* of its predecessor. (The geographic latitude and longitude of the turning point of each is shown to illustrate this fact.) The reason for this westward shift is that the saros period exceeds 6585 days by 0.32 days or about 8 hours. Thus, the earth turns from west to east by an additional 8/24 × 360° = 120°, at each eclipse in the sequence. After a triple saros period (54 years) the eclipse returns to about the same longitude. Third, the turning point gradually shifts south. (It would shift north for eclipses in a series that occurs at the descending node.) You can understand why this southward shift occurs by looking at Figure 2.5, which sketches the positions of the Sun and Moon in two successive eclipses in the series (labeled A and B, respectively). Because 19 eclipse years exceeds 223 synodic months by 0.46 days and because the Sun moves east about 60′ per day along the ecliptic, the Sun will return to a position "B" on the ecliptic 28′ west of its former position (0.46 days × 60′/day = 28′). Thus, at the second eclipse, the Moon's disk center lies further south than at the first eclipse, and as a result, the axis of the Moon's penumbral shadow will touch the Earth at a more southerly latitude.

Since each eclipse in the series occurs 28′ west of its predecessor, the last eclipse will fall just inside the western ecliptic limit, and the first just inside the eastern limit. We can now estimate how many times an eclipse can repeat in a saros series. Let's assume that the average ecliptic limit is about the average of the major and minor limits, or 16° 26′. We find that we can fit 2 × 16° 26′/28′ = 70 eclipses, spaced at intervals of 28′, within the double "average" ecliptic limits. More refined calculations give *73* eclipses per series, on the average. Thus the average series lasts 73 × 18.03 = 1315 years. The series shown in Table 2.1 will end in the partial eclipse of March 21, 2452.

A saros series begins with a group of partial eclipses, visible at high geographic latitudes, followed by a group of annular or total eclipses, at mid-geographic latitudes. Finally, the series ends with a group of partial eclipses near the opposite geographic pole.

Every eclipse belongs to such a saros series. However, the eclipses within any given year obviously belong to different saros series, since they follow at intervals of a month or 6 months, not 18 years. In 1878, T. R. Oppolzer published his monumental *Canon of Eclipses,* which contains the calculated dates, paths and other detailed information for 8000 solar and 5200 lunar eclipses that occur between 1207 B.C. and A.D. 2161. Oppolzer found that eclipses occur at an average rate of 238 per century, of which 84 (35%) are partial, 77 (32%) are annular, 11 (5%) are combinations of total and annular, and 66 (28%) are total. Thus, 42 solar eclipses occur, on the average, in a saros period of 18 years, 11 days. This means that at any time, 42 *different saros series* must be running in parallel!

G. Van den Bergh studied Oppolzer's eclipse calculations intensively and found several other regularities in the appearance of eclipses. (See his *Periodicity and Variations of Solar and Lunar Eclipses,* T. Jeenk Willink and Zoon N. V., Haarlem, 1955.) Among the most important is the existence of another period, the "inex," which lasts 358 synodic months or 29 years minus 20 days. Like the saros, the inex is a "beat period" between two nearly equal natural periods: 358 synodic months equals 388.5011 draconic months. Thus an eclipse at, say, the ascending node will repeat at the descending node (because of the 0.5 draconic month difference) after 358 new moons. The remaining small inequality of the two intervals (.00011 draconic months or 4.3 minutes) shifts the second eclipse by a mere .0411° east with respect to the descending node.

The eclipses in an inex series are visible alternately at symmetric geographic latitudes. The series starts with 140 constantly increasing partial eclipses that move progressively from the poles toward the equator. Next follow 250 central eclipses that move toward the equator, another 250 central eclipses that move back toward the poles and a final 140 partials that move toward the equator.

Each eclipse in the series shifts .0411° east with respect to a node. After 780 eclipses, the cumulative shift equals the double (average) ecliptic limits—about 34°—and the series ends.

Each inex series lasts, therefore, 780 × 29 = 23,000 years! Since the eclipses in the series follow at intervals of 29 years, and 238 eclipses occur per century, on the average, about 70 inex series are running in parallel at any moment.

Van den Bergh rediscovered the important fact that a saros series (containing 73 eclipses at intervals of 18 years and lasting 1315 years) is "born" every inex, i.e. every 29 years. That is, the first eclipse in the saros series (a partial eclipse) occurs at the eastern eclipse limit. Thus, *every* eclipse can be assigned a place in a saros series and *every* saros a place in an inex series. Van den Bergh constructed a beautiful "panorama of eclipses," a grid in which all of Oppolzer's eclipses appear. Each column of the grid is a saros series, each row an inex series.

Although the saros interval is well known, the inex interval is not. Van den Bergh found only two references to it in the scientific literature of the last century: a "discovery" paper by J. N. Stockwell in 1901 and a reference to the double inex period by J. T. Menendez in 1941.

The saros and inex give astronomers a marvelous working tool for predicting eclipses both in the future and in the past. However, with high speed digital computers and with an accurate mathematical description of the Moon's motion, astronomers cannot calculate all the desired circumstances of eclipses in marvelous detail from first principles. You might enjoy browsing through Oppolzer's *Canon of Eclipses,* which was reprinted by Dover Publications in 1962. The book has a good collection of the maps of centerlines of coming eclipses, which will help you to plan your travel for the next 300 years. The book is also useful for establishing historical dates, as we saw in Chapter 1.

Oppolzer's massive work has been updated by J. Meeus, C. C. Grosjean, and W. Vanderleen in the *Canon of Solar Eclipses.* This huge tome contains the elements of all solar eclipses taking place between July 1898 and March 2510. In addition to a fine collection of maps, these authors give detailed tables of the geographical coordinates of each center line, the width of the eclipse path and the duration.

For more detailed planning, you should get acquainted with the *Astronomical Almanac,* formerly titled *The American Ephemeris and Nautical Almanac,* which is compiled by the U. S. Naval Observatory and published by the U. S. Government Printing Office. The *Astronomical Almanac* appears at least one year in advance and contains preliminary information on all the solar and lunar eclipses that will occur in that coming year. For each eclipse, there is a map of the path. The *Almanac* contains a table of the coordinates of the center line, the universal time of the maximum phase of the eclipse, whether it be total or annular, the duration at various points along the center line, and the northern and southern limits of the eclipse path.

If you plan to make astronomical observations, the *Almanac* gives you the "elements"—that is, the astronomical coordinates of the Sun and Moon and their hourly motions. A few months before an important eclipse, the U. S. Naval Observatory will issue a circular giving the local circumstances of the eclipse for many points along the center line. The exact time of maximum phase, the configurations of the Sun and Moon, the eclipse magnitude, as well as the coordinates of the second and third contact points are given. If you *really* want to get into details, you should read the explanatory supplement to the *Astronomical Almanac,* which shows you how to make your own predictions of the circumstances of an eclipse for any location from the basic data that the *Almanac* provides.

In addition to all this, the Naval Observatory predicts the profile of the Moon's edges, and therefore the sequence of Bailey's Beads that will be seen under perfect circumstances. It is a pity that meteorologists cannot give us anywhere near as accurate a forecast of the weather that will prevail during a coming eclipse. But, except for the weather, you can plan everything else for your next eclipse expedition.

NOTES

1. From analytic geometry we learn that an ellipse is the path of a point which moves so that the sum of its distances from two fixed points, which are called foci, is a constant. Kepler's First Law states that planets move in ellipses with the Sun as one of their foci. A circle is an ellipse in which the two foci overlap and the major and minor axes of the ellipse are equal.

2. The Earth completes a revolution about the Sun in 365.25 days. As a reflex of this motion, the Sun appears to move against the background of the distant stars, along a track called the ecliptic, at 360°/365.25 days = 1°/day, approximately. Similarly, the Moon completes a revolution about the Earth in 27.32 days (the sidereal month), with an average eastward motion against the distant stars of 360°/27.32 = 13.2°/day.

   The "day" we are referring to here is the mean solar day in the astronomer's jargon. As the Earth rotates, the Sun appears to rise and set. "Local noon" occurs when the Sun lies due south of your position (if you are in the Northern Hemisphere, or due north in the Southern Hemisphere). The interval between two successive local noons, averaged throughout the year, is a mean solar day. This is the day we normally deal with in everyday life.

3. Kepler's Second Law describes how the angular motion of the Moon varies according to its distance from the Earth. Imagine an elliptical lunar orbit, with the Earth at one focus. Now, imagine a line drawn between the centers of the Earth and Moon. Kepler's law states that this line will sweep out equal areas in equal time intervals. When the Moon is near perigee, the line is short, so its angular speed is rapid. Near apogee, the line is long, so the Moon's angular speed is slower.

4. You may wonder why the synodic and sidereal months differ. One sidereal month (27 days) after a new Moon, the Moon has returned to the same point in its orbit. However, the Earth has revolved in its orbit by 1° per day or about 27° during this time. Hence the Moon must revolve an additional 27° (taking 2 days) before it once again lies between Earth and Sun, i.e. until new moon. More precisely, we can say that the number of new moons per year (N) equals the difference between the number of lunar revolutions per year (365.25/27.32) and the number of Earth's revolutions per year (1.0). Thus N = 365.25 days/27.32 days − 1. So N = 12.369 and the synodic month is 365.25/12.369 = 29.53 days.

SUGGESTIONS FOR FURTHER READING

Books

Meeus, J., Grosjean, C. C., and Van Der Leen, W. *Canon of Solar Eclipses.* Pergamon Press: 1966.

Oppolzer, T. R. *Canon of Eclipses.* Dover Publications: 1962.

Pasachoff, J. M. *Contemporary Astronomy,* W. B. Saunders Co: 1977.

Van Den Bergh, G., Willink, T. J., and Zoon, N. V. *Periodicity and Variations of Solar and Lunar Eclipses.* Haarlem: 1955.

3
Time, Tide, and Gravity

In 1693, Edmund Halley, fresh from the discovery of his now famous comet, turned to a study of the Moon's motion. He analyzed a group of lunar eclipses recorded in Ptolemy's second century *Almagest,* (a compendium of ancient Greek astronomical observations), a group of ninth century Arabian solar eclipses, and a group of seventeenth century solar eclipses. He concluded that the Moon's mean speed in its orbit was *not* constant, but had been accelerating throughout historical time. Now, Kepler's second law (see Note 3, Chapter 2) leads us to expect that the *instantaneous* speed of the Moon will vary according to its distance from the Earth, but when averaged over many complete orbits, the *mean* speed, by definition, should be constant. Halley's discovery indicated that something was seriously wrong with gravitational theory, astronomical observations, or both.

Richard Dunthorne was the first to estimate the size of this so-called "secular acceleration" of the Moon. In 1749, he tried to reconcile the dates of five eclipses: a lunar eclipse observed in Babylon in 721 B.C., Theon's eclipse in A.D. 364, and one lunar and two solar eclipses dating to the tenth century. He concluded that the Moon must have accelerated an additional 10 arcseconds every century, or in other words, the secular acceleration amounts to 10 arcseconds per century per century.

Laplace, Kepler, and other mathematical luminaries of the eighteenth century proposed explanations for this curious anomaly. The proposed causes, then as now, fall into two basic catagories: *gravitational* perturbations of the Moon by the Sun and planets, and *nongravitational* perturbations of the Earth-Moon system, by such forces as the friction of the ocean tides.

This discrepancy in the motion of the Moon may seem esoteric, or trivial, or both. It has persisted to the present day as an important astronomical problem, however, for several reasons. First, as we shall see later in this chapter, the motion of the Moon is still used to determine astronomical time, so there is a very practical reason for wanting to understand the Moon's acceleration. Second, geophysicists are intrigued with the acceleration because it gives an indirect estimate of the friction of ocean tides, which they wish to measure and compare with theory. Third, tidal friction in the Earth's crust, redistribution of material in the Earth's interior, and the coupling of the Earth's core to its mantle could each change its spin properties in such a way as to affect the Moon's motion. If these "nontidal" terrestrial effects could be detected from the Moon's motion, geophysics would benefit accordingly.

What is more, P. A. M. Dirac suggested in 1937 that the coefficient (G) in Newton's law of gravitation, usually called the "gravitational constant," may actually vary over long time spans. Several rival theories of general relativity also predict such a change. The Moon's motion would be very sensitive to a change in G, and the lunar acceleration could offer a direct test of cosmological theories.

Finally, the secular acceleration offers a test of the consistency and reliability of historical records of lunar and solar eclipses. These records have value for dating events in the history of ancient civilizations. Critical examinations of these records reveal disturbing discrepancies, some of which are the result of pure invention, as we saw in Chapter 1.

DYNAMICS OF THE EARTH-MOON SYSTEM

The Earth and Moon are bound by their gravitational attraction. Newton's gravitational law states that the attractive force, F, is proportional to each of their masses (M_e and M_m) and varies inversely as the square of their separation, r. Thus:

$$F = \frac{G M_e M_m}{r^2}$$

where G is the gravitational "constant." The force of attraction, F, on the moon balances at every moment the centrifugal force, $\omega^2 r M_m$, produced by its orbital motion. (Here ω is the angular speed of the

Moon in degrees per second or radians per second). The same is true for the Earth. As a result, the Earth and Moon revolve in elliptical orbits about a special point, the center of gravity of the system. (See Figure 3.1.) The center of gravity lies on a line joining the centers of the two bodies. Its distance from each body (r_e and r_m) is inversely proportional to the body's mass; that is:

$$r_e / r_m = M_m / M_e.$$

Each body possesses a physical property called its orbital angular momentum which measures the tendency of the body to continue in its orbital path. The angular momentum of the Moon with respect to the center of gravity of the Earth-Moon system is $L_m = M_m \omega r_m^2$. The total orbital angular momentum of the Earth and Moon is just the sum of their individual momenta:

$$L_o = L_e + L_m.$$

The Earth's spin also creates angular momentum, L_s, so the total for the system is:

$$L_o + L_s = L_T.$$

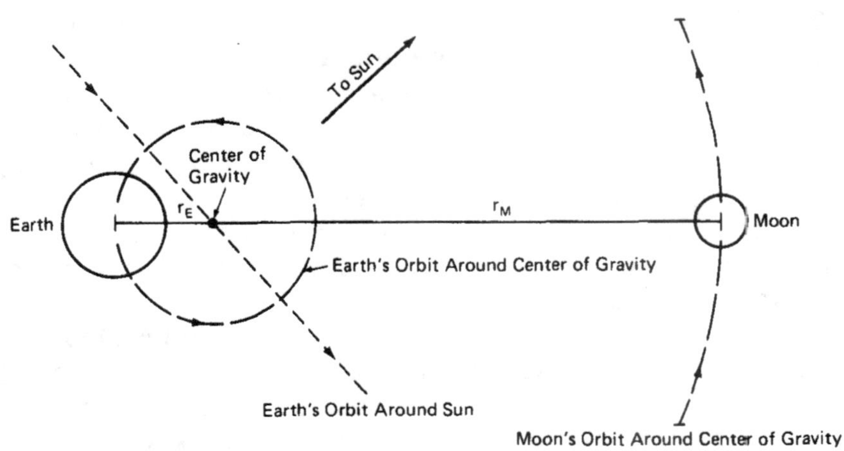

Fig. 3.1. The Earth and Moon revolve about a common center of gravity. The distance of each body (r_e and r_m) from this center is inversely proportional to the body's mass.

The orbital momenta represent about 80% of the total. The total system momentum (L_T) is important because it will not change throughout all of time, no matter how complicated the motions of the Earth and Moon may be, if we neglect small perturbations by the Sun and planets. All that can happen is that the individual momenta (L_s, L_e, and L_m) may change, but their *sum* (L_T) will remain constant. Let us see how the Moon's momentum may change.

The Moon's attraction raises a tidal bulge in the Earth's oceans. If you've ever watched the tides in a harbor or at a beach, you can appreciate the massive flow of water that is involved. In shallow seas and on the margins of continents, this tidal flow exerts a frictional force which tends to slow down the Earth's spin and lengthen the day. (Immanuel Kant, better known as a philosopher, first suggested this effect in 1754). As the Earth's spin decreases, so does its spin angular momentum. In order to conserve the total angular momentum of the Earth-Moon system (L_T), the Moon's orbital momentum (L_m) must increase. As a result of Kepler's Second law and Newton's Gravitational law, the Moon recedes from the Earth and slows down; that is, the sidereal month lengthens.

It is important for what follows that you understand that tidal friction *slows down* the Moon. The Earth's spin, however, is used as a clock, and if this clock is running slow, the Moon may appear to speed up or *accelerate* in its orbit. Then, the Moon's angular position would be ahead of the one we would predict from the laws of mechanics.

Tidal friction is a nongravitational effect, and the explanation for the lunar acceleration that we have just sketched is an example of a whole class of such effects. If the Earth's spin slows down for *any* reason, such as the restructuring of its core or the coupling of its core and mantle, the Moon will respond by receding and slowing down, just as though tidal friction were at work.

In contrast, there are strictly gravitational effects that can produce a lunar acceleration. The simplest of these was proposed by Laplace, and runs as follows.

The Sun's gravitational pull perturbs the motion of the Moon, by an amount that depends upon the mean distance between the Sun and the Moon, or equivalently, between the Sun and the Earth. The eccentricity of the Earth's orbit has been decreasing because of planetary perturbations since the creation of the solar system. As a result, the mean distance between the Sun and the Moon has increased, and the net effect of this is to shorten the sidereal month. Now if we reject the possibility that the Earth is slowing down, as Laplace did, we conclude that the Sun speeds up the Moon, i.e. produces a *positive* lunar acceleration. Laplace calculated a value that agreed fairly well with Dunthorne's estimate of + 10″ per century per century. The effect is certainly real, and has the opposite direction from that of tidal friction which slows down the Moon. E. W. Brown, in 1920, estimated the solar effect at 7.14″ per century per century.

THE ECLIPSE METHOD

The Moon's acceleration is exceedingly small: a mere 10″/century/century. Special methods, based on eclipses of stars by the Moon, or laser ranging, enable astronomers to locate the Moon with a precision of perhaps .01″ so that after only a few years of collecting data they can determine the lunar acceleration with high accuracy. Of course, high precision in time measurements is also needed for either of these methods to work, because the acceleration corresponds to only about 0.2 seconds of time per year. Atomic clocks easily provide the necessary accuracy and stability. We shall return to these methods, and another, later on.

However, these techniques determine the *present* acceleration, and yield no information on its past behavior. For this reason, the analysis of ancient solar (and lunar) eclipses that span many centuries continues to be useful. Eclipses have another remarkable advantage. As R. R. Newton wrote in 1969:

> An observation that a solar eclipse was total at a known place is an observation of rather high precision. It is not necessary for the observer to record the time accurately. In almost all cases, we need the time only within a *decade* in order to identify an eclipse; the place then determines the time with an error of only a few minutes. Thus a person with no astronomical training can make a valuable astronomical observation.

Since the Moon's acceleration is gravitationally coupled to the Earth's rotation, and since the Earth's rotation has been used throughout history as the primary standard, the astronomer's task in analyzing ancient eclipses is especially difficult. He must determine *both* the rate of the Earth's spindown and the lunar acceleration simultaneously from the eclipse data.

Perhaps the most widely quoted of such investigations of the lunar acceleration is that of J. K. Fotheringham, published in 1920. Fotheringham used the times and magnitudes of ancient lunar eclipses, lunar occultations of stars, equinox observations, and observations of solar eclipses in order to determine the acceleration of the Moon and Sun. The acceleration of the Sun is only apparent, of course, and simply reflects the spindown of the Earth.

Fotheringham selected eleven solar eclipses for his study. They cover a period of 1400 years, ending with the eclipse of Theon in 364 A.D. This eclipse is especially important because the astronomer Theon carefully noted the local times of the beginning, middle, and end of the total eclipse. For all the other eclipses, Fotheringham knew only their location, whether they were total or annular, and their approximate dates.

Fotheringham's method is typical, in principle, of all later work, and he describes it in clear language:

> I adopt elements of the motion of the Sun and Moon supposed to be approximately correct, and compute the belt of totality or annularity for each eclipse. I then make similar sets of computations with two sets of elements, in which the secular accelerations of the Moon and Sun respectively have been increased by 1″ per century... From the belts of totality I compute equations showing the limits between which the two accelerations must lie in order to satisfy the presumptions made in regard to each eclipse. . . Graphs representing the different equations are then plotted out on a sheet of paper, and it is thus possible to see to what extent the presumptions are consistent and what values of the two accelerations will satisfy the presumptions which are found to be consistent.

Fotheringham found the three most consistent eclipses are those mentioned by Hipparchus (130 B.C.), by Plutarch (71 A.D.), and in the Assyrian chronicles (763 B.C.) The eclipse records of Theon, although apparently accurate, required radically different solar and lunar accelerations than all the other eclipses he selected and Fotheringham chose to ignore it. Fotheringham found the solar and lunar accelerations were +1.5 and +10.8 arcseconds per century per century, respectively, when referred to mean solar time. (See Table 3.1.) Remember that mean solar time uses the Earth's rotation as a clock, and the Sun as a marker. If we assume the rotation is uniform for the purpose of timekeeping, the Moon will seem to advance *ahead* of its predicted position, so that its acceleration is *positive*.

Fotheringham's work has been heavily criticized by recent investigators, especially by R. R. Newton, on two counts: (1) some of the eclipses he selected never occurred, e.g. Plutarch's, and (2) he played the "identification game." According to Newton, the game involves adopting a preliminary estimate for the lunar acceleration, predicting the circumstances of ancient eclipses and then discarding those eclipses that disagree too strongly with the prediction! The calculations are then repeated with "reliable" observations and yield, not surprisingly, values that differ little from those assumed initially. Newton considers that most investigators of the twentieth century have played this game. Eclipse records, say Newton, should be selected in advance of any calculations, for their historical authenticity, and discarded only if they fail the test.

In 1939, H. Spencer-Jones determined the solar and lunar accelera-

Table 3.1 The Moon's Secular Acceleration from Solar Eclipses (Arcseconds per Century per Century)

AUTHOR	DATE	EPOCH	ACCELERATION
Dunthorne	1749		10.8*
Adams	1853		8.3*
Airy	1857		12–13*
Newcomb	1878		8.8*
Ginzel	1883		11.47*
Fotheringham	1920	200 B.C.	−30.8
De Sitter	1927	200 B.C.	−37.7 ± 4.3
Spencer-Jones	1939	1800 A.D.	−22.4 ± 1.1
Newton	1970	200 B.C.	−41.6 ± 4.3
		1000 A.D.	−42.3 ± 6.1
Stephenson	1972	300 B.C.	−34.2 ± 1.9
Muller	1975	470 B.C.	−34.5 ± 3.0
Muller, Stephenson	1975	400 B.C.	−37.5 ± 5.0
Muller	1978	1375 B.C. to 1567 A.D.	−30.3 ± 3.0
Newton	1979	1800 A.D.	−28.4 ± 5.7
		200 B.C.	−31.57 ± 8.98

* These values refer to mean solar time, the remainder to ephemeris time. (See text for explanation.)

tions from astronomical data, *excluding* solar eclipses, and found quite different results from everyone before him. He analyzed fluctuations in the positions of the Sun, Mercury, Venus, and the Moon. He used observations going back to 1681 for the Moon, 1761 for the Sun, and 1678 for Mercury, i.e. all relatively recent measurements. He found solar and lunar accelerations of 1.07 ± .06″ and 3.11 ± .57″/century/century when referred to mean solar time. Jones concluded:

> These values of the acceleration will not satisfy any of the ancient observations of eclipses and occultations, which on the whole are in very good agreement with one another in requiring appreciably larger values. There seems to be no escape from the conclusion that the effects of tidal friction are appreciably less at the present time than the average effects over the past two thousand years.

Spencer-Jones' work was tremendously exciting to astronomers and geophysicists. First, it showed that the orbital motions of the planets are consistent, and serve as a better clock than the Earth's spin or the Moon's motion. A new astronomical time-scale, "ephemeris time," was introduced to replace mean solar time. Ephemeris time, in principle, is the time that appears in the equations of orbital motion of the planets. In practice, it is determined from the Moon's orbital motion, which is corrected by Spencer-Jones' standard value for the lunar acceleration due to tidal friction. When referred to ephemeris time, this value is −22.44″ ± 1.1″/century/century. Secondly, Spencer-Jones result suggested recent changes in the strength of tidal friction, which intrigued geophysicists. He stimulated further efforts to improve determinations of the lunar acceleration from solar eclipses and to decide whether it has changed during historical times.

Table 3.1 summarizes the results of investigations following Spencer-Jones, and Figure 3.2 displays them for easier comparison. The "epoch" refers to the mean date of the observations analyzed. Notice that only three authors have picked up Spencer-Jones challenge (Stephenson, Muller, and Newton), but they have all worked diligently. We saw their independent selections of "reliable" eclipses listed in Table 1.1. They agree moderately well on an average lunar acceleration around −35″/century/century, or −33″/century/century if we drop Newton's large 1970 figures. In contrast stands Spencer-Jones "standard" value of −22.4″/century/century: a clear discrepancy. Newton reworked Spencer-Jones' material in 1979 and improved somewhat the comparison with eclipses, but a significant discrepancy still exists.

Does the lunar acceleration change during historical time? Perhaps a slight trend can be discerned in Figure 3.2, but the scatter and uncertainty of the numbers leaves the question open. Moreover, the eclipse method disagrees with recent noneclipse methods, which we will describe next.

Incidentally, the eclipse solutions also yield the rate of spindown for the Earth. The day increases in length by about 2 milliseconds per century. This figure may convey, more effectively than Table 3.1., how small are the effects being considered!

NONECLIPSE DETERMINATIONS OF THE MOON'S ACCELERATION

Let us turn next to three special methods that yield contemporary values of the lunar acceleration and which don't depend on an analysis of eclipses.

The first method is laser ranging of the Moon. During the Apollo program of the 1970s, the American astronauts placed several retroreflectors on the Moon's surface. These "corner cubes" have the pro-

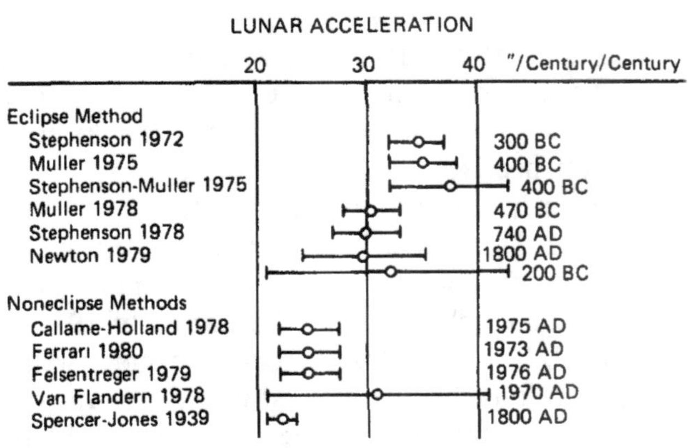

Fig. 3.2. Determinations of the lunar secular acceleration by different investigators and by different methods. The dates shown refer to the average date of the data.

perty of returning a beam of light exactly in the direction from which it comes. A group of physicists from Princeton University and the University of Texas have been measuring the Moon's position, using these retro-reflectors ever since then. They direct an extremely short pulse of laser light at the moon and time its round-trip travel time with an atomic clock. Knowing the speed of light, they can then determine the distance of each reflector within a few tens of centimeters and, repeating the experiment, its speed of approach or recession. These quantities serve to locate the Moon extremely precisely in its orbit, and eventually yield the lunar acceleration. Note that *distances* and *speeds* are measured, not angles.

In 1978, O. Calame and J. D. Mulholland reported on seven years of such observations. Their value for the lunar acceleration (−24.6 ± 1.6″ per century per century) refers to atomic time, not ephemeris time. More recently, Ferrari and coworkers have confirmed this result. During 1969 through 1977, Lunar Orbiter 4 circled the Moon in a polar orbit, every 12 hours. The telemetry signals from the Orbiter were analyzed to give its instantaneous speed toward or away from the Earth. At the same time, the distance of the Moon was determined by laser ranging. Together, the two sets of data give precise distances and radial speeds of the Moon, on an atomic time scale. Ferrari's result (−23.8 ± 3.1″/century/century) agrees with other atomic time determinations, but disagrees with the eclipse results.

The Moon eclipses or "occults" stars as it moves across the sky. Because the Moon has no atmosphere, the eclipse occurs very abruptly. The instant at which an occultation occurs can be recorded accurately and this measurement is unaffected by scintillation in the Earth's atmosphere, in contrast to an angular measurement. By recording the times of occultations of stars whose positions on the sky are accurately known (from many precisely measured photographs, for example) an astronomer can determine the Moon's motion and eventually its secular acceleration. This method has been used effectively in the past to determine the secular accelerations, but depended until 1955 on suspect astronomical clocks, such as the Earth. In 1975, T. C. van Flandern, of the U. S. Naval Observatory, analyzed twenty years of occultation observations that were timed with a cesium atomic clock. He found a *huge* value for the lunar acceleration: −65″ ± 18″/century/century which he later revised to −36″.0 ± 5.0″/century/century. We will come back to his interpretation a little later in this chapter, after we consider yet another technique.

In recent years, geophysicists have been able to determine accurately the frictional force on the Earth produced by the ocean tides, and then to estimate the rates at which the Earth's rotation and Moon's orbital speed are decelerating. The physical principle of the method is easy to grasp, but the mathematics are formidable.

If the Earth were a perfectly smooth rotating ball, covered uniformly with water, the Moon's attraction would raise ocean tides that heap up directly opposite the Moon, as shown in Figure 3.3. The Earth's continents impede the flow of the ocean water, however. As a result, the tidal "hump" leads the Moon by a small angle, as shown in Figure 3.3. This angle, and its time variations, are indirect measures of the frictional force the ocean tides exert on the Earth.

An artificial satellite, in orbit near the Earth, will experience the combined gravitational attractions of the Earth, the Moon, and the ocean tidal hump. Of all the orbital parameters, the inclination of the satellite orbit to the Earth's equatorial plane is most sensitive to these perturbations. But the effects are tiny: the angle of inclination typically varies by only .04 arcseconds in a period of 10 to 20 days. Nevertheless, by accurately tracking such satellites, geophysicists can determine the mass distribution in the ocean tides, and therefore the lead angle shown in Figure 3.3. This angle then leads to an estimate of the lunar acceleration. Some recent results, due to T. L. Felsentreger and J. G. Marsh, are compared in Figure 3.2 with the values obtained with other techniques. The method agrees fairly well with laser ranging of the Moon, and with good old reliable Spencer-Jones, but again conflicts with the eclipse results!

VARIATION OF THE GRAVITATIONAL CONSTANT

We saw earlier that tidal friction can explain the lunar acceleration. Let us turn now to another physical factor that may influence the

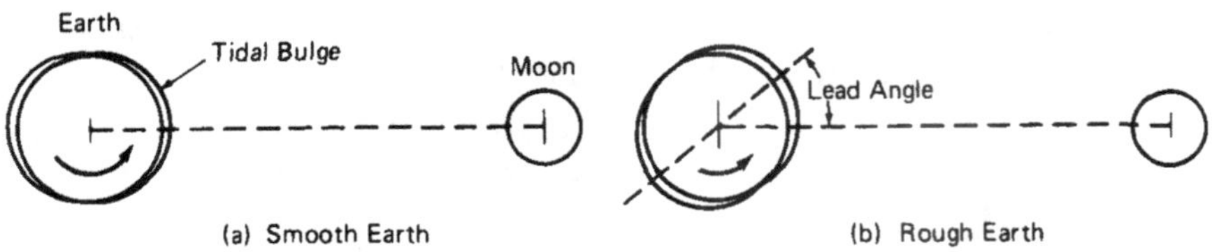

Fig. 3.3. The Moon raises ocean tides on the Earth. If the Earth were smooth the tidal bulges would line up with the Moon—as in (a)—but since the Earth is rough the bulges lead the Earth-Moon line by a small angle—as shown in (b).

Moon's real or apparent motion. In 1937, P. A. M. Dirac noticed a remarkable coincidence between two large, dimensionless numbers that arise in physics. A proton in a hydrogen atom attracts its electron both electrostatically and gravitationally. The electrical attraction exceeds the gravitational by a factor of 10^{40} (one, followed by forty zeros). On the other hand, the ratio of the age of the universe and the period of revolution of an electron in hydrogen atom is also about 10^{40}. Dirac suggested that these two huge numbers agree for good, physical reasons and not simply by accident. He showed, very simply, that the two numbers would agree *exactly* if Isaac Newton's gravitational "constant" (G) were *decreasing* at a rate proportional to the time. Dirac estimated in 1973 that G decreases by 6 parts in 100 billion each year, and at that rate would have halved during the present age of the universe. Such a decrease would affect theories of cosmology tremendously. In particular, a decreasing G could account for the expansion of the universe.

Now Einstein's theory of general relativity predicts no such change in G. Three theories, other than Dirac's, do: (1) that of F. Hoyle and J. Narlikar, (2) that of C. Brans and R. H. Dicke, and (3) that of V. Canuto and his colleagues. Several geophysical and astrophysical tests of these theories are possible; we shall consider here only the secular lunar acceleration.

As we saw earlier, Van Flandern obtained a value for the lunar acceleration of $-65 \pm 18''$/century/century, by using an atomic clock to time occultations of stars by the Moon. He proposed to account for the large difference between his result and the average value from four eclipse solutions ($-38'' \pm 4$) by invoking a slow change in the gravitational constant. Remember that his value refers to the uniform time measured by an atomic clock, whereas the eclipse results refer to ephemeris time, which the orbital motions of the Earth, Moon and planets obey. If G were to decrease, at a rate of $\mathring{G}/G$ parts per year, the solar system would expand. To conserve angular momentum, the mean orbital speed of a planet would then decrease by $\mathring{\omega}/\omega$ parts per year, where the two rates relate as follows:

$$\mathring{\omega}/\omega = 2\mathring{G}/G.$$

This statement is equivalent to saying that a clock that keeps ephemeris time slows down at a rate of $2\mathring{G}/G$ parts per year, relative to an atomic clock.

The Moon's acceleration, measured with an atomic clock, includes the effects of tidal friction, the change in G, and perhaps other effects. The acceleration measured with an "ephemeris" clock excludes the G effect, which is absorbed in the slowing down of the clock. The difference in the secular acceleration of the Moon, as measured on these two time scales, should be a direct measure of the rate of change of G. Van Flandern therefore subtracted the eclipse value for the lunar acceleration from his own and found $\dot{G}/G = -8 \pm 5$ parts per 100 billion per year. When he later revised his own figure downward (to 36″/century/century), however, the result turned out to be insignificant: $\dot{G}/G = -2 \pm 2$.

P. M. Muller, one of that indefatigable trio that continues to analyze ancient eclipses, has attempted to test several theories of general relativity that predict changes in G. (See his paper "On the Measurement of Cosmological Variations of the Gravitational Constant," L. Halpern, editor, U. Florida Press, 1978). He first determined the lunar and solar accelerations from eclipses, from transits of Mercury across the solar disk, and from the observed times of solar equinox (see Glossary). Then he solved simultaneously for the tidal friction, nontidal friction (which turns out to be negligible), and for $\dot{G}/G$. Each theory of gravitation provides a different relationship among these quantities and the Moon's acceleration, and by inserting the empirical values in each equation, Muller was able to test each theory independently. His results are shown in Table 3.2.

Muller concluded that (1) there is strong observational support for a significant change in G during the life of the Universe, and (2) both the Dirac and the Brans-Dicke theory survive the test.

Since Muller completed this work (1978), the Brans-Dicke theory

Table 3.2 Comparison of Observed and Predicted Changes in the Gravitational Constant

THEORY	$\dot{G}/G$ PREDICTION	$\dot{G}/G$ OBSERVATIONAL ESTIMATE
a) Hoyle-Narlikar Dicke Peebles & Dicke	$-5.6 \pm .7 \times 10^{-11}y^{-1}$	$-2.3 \pm 1.5 \times 10^{-11}y^{-1}$
b) Brans-Dicke theory	$-\ 5\ 10^{-11}y^{-1}$	$-2.3 \pm 1.5 \times 10^{-11}y^{-1}$
c) Dirac theory	$-5.6 \pm .7 \times 10^{-11}y^{-}$	$-5.1 \pm 3.0 \times 10^{-11}y^{-1}$

has lost favor on other grounds, and at least one new theory involving $\dot{G}/G$ (Canuto's) has arisen.

SUMMARY

We have seen that Halley's discovery of the lunar acceleration has provided scientists with three centuries of "innocent merriment," leading them in directions nobody could have predicted. Let us summarize:

1. Eclipses indicate that the Earth-Moon system is winding down, mainly because of the friction of the ocean tides. The day lengthens by about 2 milliseconds per century and the sidereal lunar month by about 63 seconds per century.
2. Solar eclipse records suggest that the deceleration rates of the Earth and Moon have remained roughly constant through the past 2000 years, although there is room for argument. Fossils of organisms that are sensitive to the length of the day have been studied in order to recover the Earth's rotational history during the past 100 million years. However, these results are no more certain than those from eclipses. (The study of the Earth's rotation is a huge, lively subject that could easily fill another book!)
3. Ephemeris time is about the most accurate that astronomers can provide, but it suffers from residual uncertainties in the Moon's and Earth's deceleration. An atomic clock must be used if an accuracy of one part in 100 billion for a year is needed.
4. However, the lunar acceleration does seem to depend on whether ephemeris time (as in the eclipse method) or atomic time (as in other methods) is used to measure it. (See Figure 3.2.) This difference can be interpreted as evidence for a decreasing gravitational constant.
5. Accurate estimates of the rate of change of G must probably await another decade of laser ranging of the Moon. Preliminary estimates, derived from a comparison of the lunar accelerations measured on ephemeris and atomic time (Van Flandern's method), suggest that G varies by a few parts in 100 billion per year. Muller says that Dirac's cosmological theory seems to predict correctly the observed change in G, unlike the "primitive" cosmological

theories. However, the "observed" change in G is probably still too uncertain to discriminate unambiguously among rival theories.

SUGGESTIONS FOR FURTHER READING

Books

Halpren, L. (ed.) *On the Measurement of Cosmological Variations of the Gravitational Constant*. University of Florida Press: 1978.

Newton, R. R. *The Moon's Acceleration and Its Physical Origins*. Vol. I Johns Hopkins University Press: 1979.

Periodicals

Calame, O., Mullholland, J. D. *Science* **199**, 977, 1978.

Felsentreger, T. L., and Marsh, J. G. *Journal of Geophysical Research* **84**, 4675, 1979.

Ferrari, A., *et al. Journal of Geophysical Research* **85**, 3939, 1980.

Fotheringham, J. K. *Monthly Notices of the Royal Astronomical Society* **81**, 104, 1920.

Newton, R. R. *Science* **166**, 825, 1969.

Spencer-Jones, H. *Monthly Notices of the Royal Astronomical Society* **99**, 541, 1939.

Van Flandern, T. C. *Monthly Notices of the Royal Astronomical Society* **170**, 333, 1975.

Wesson, P. S. "Does Gravity Change with Time?" in *Physics Today*. July 1980, p. 32.

4
The Chromosphere

INTRODUCTION

Just beyond the sharp edge of the visible Sun, the chromosphere begins. Invisible in full daylight, the chromosphere blazes out in red glory for a few seconds (at second and third contacts) during a total eclipse. The chromosphere is a strange region that lies between the cool photosphere and very hot corona. It is the lowest region of the Sun's atmosphere in which dynamic and magnetic processes, as well as radiation, heat the gas. Shock waves form here from sound waves that have traveled upward from the noisy photosphere. Heat flows downward from the corona, is converted to light in the chromosphere, and is radiated into space. Magnetic fields, twisted by photospheric motions, transport hydromagnetic waves into the chromosphere, where they dump their energy.

Solar astronomers have been fascinated by the strange phenomena in the chromosphere ever since its discovery. In Chapter 1, we reviewed some of the work of the pioneers in the field, and in this chapter we will pick up a small part of the story. Remember that total eclipses are no longer the only or even the best means of studying the chromosphere. Observations with coronagraphs (out of eclipse), with special narrow-band filters that reveal the chromosphere on the disk of the Sun and with rocket and satellite-borne spectrographs that view the chromosphere at extreme ultra-violet wavelengths (1000 to 3000 Angstroms) have all taken their place as important modern techniques. Nevertheless, total eclipses continue to draw professional astronomers who want to unravel the strange physics of the chromosphere.

Why is this true? There are two main reasons. First, the entire chromosphere is only 5000 to 10,000 km thick, or about 0.4% of the solar diameter. Astronomers need observations at close intervals (say 200 or 300 km apart), since the temperature and density change rapidly with increasing height. Now even the best solar telescopes, at the best locations, rarely resolve solar details smaller than about 500 km because of scintillations in the Earth's atmosphere ("seeing"). However, during a total eclipse, the Moon acts as a sharp, black shutter that covers (or uncovers) the Sun at a uniform speed that corresponds to 300 or 350 km/s in the chromosphere. Thus, if an astronomer can take photographs or spectra or simple flux measurements rapidly enough, the Moon's motion will provide the spatial resolution he needs.

The second reason also has to do with spatial resolution. As we shall see, radio astronomers want to measure the brightness distribution of the chromosphere at wavelengths near 1 millimeter. The resolving power of a telescope decreases in direct proportion to the wavelength of light used in the observation. At 1 mm, even the 36*foot* telescope at Kitt Peak National Observatory resolves solar details no smaller than 40 arcseconds or about 30,000 km. Thus, all details within the chromosphere are smeared out by the "instrumental" resolution. Only a few millimeter wave interferometers can resolve the chromosphere. Once again, the Moon's motion at eclipse gives radio astronomers the spatial resolution they need.

THE CHROMOSPHERE IN VISIBLE LIGHT

The Jumping-film Era

In 1950, three famous institutions (the High Altitude Observatory, the Sacramento Peak Observatory, and the National Bureau of Standards) resolved to field a whole series of eclipse expeditions during the following decade to record and analyze the flash spectrum of the chromosphere. They were inspired by the observational work of D. H. Menzel of Harvard University who used his "jumping-film" spectrograph to photograph the chromospheric spectrum at the eclipse of 1932 and 1936 with unprecedented speed and completeness. You will remember from Chapter 1 that Menzel's spectrographs had no entrance slit to define the width of the spectrum lines, but used the narrow chromospheric crescent itself as a luminous slit source. Also, the key feature of the spectrograph was its camera, which could snap off a complete spectrum every second as the Moon advanced. R. N. Thomas and R. G. Athay, collaboraters of Menzel, pointed out that no further progress could be made in understanding the nonthermal heating of the chromosphere until its temperature and density had been determined from the analysis of chromospheric spectra, and that existing observations were simply inadequate. Thus, the three institutions agreed to cooperate in a decade-long effort.

They built a new version of the spectrograph, with three cameras to cover the wavelength regions 3360 to 5100 A, 3950 to 6500 A, and 6200 to 8680 A. J. W. Evans led an expedition to Khartoum, in the Sudan, to observe the eclipse of February 25, 1952. Despite the incredible difficulties in keeping their complicated machine working in the midst of a desert, they succeeded in photographing a beautiful set of spectra. One spectrum was taken every 220 km (or 0.6 seconds) up to a height of 3000 km above the solar limb, and every 5000 km thereafter up to a height of 8000 km. Their spectra were similar to that shown in Figure 4.3 (see page 61).

In addition to hydrogen, helium, and metallic spectrum lines, the cameras recorded the intensity of the continuous spectrum between the lines. R. G. Athay, D. H. Menzel, J. C. Pecker, and R. N. Thomas showed how to derive the temperature and density of the chromosphere from the wavelength variation of the continuum intensity. The continuous light that the chromosphere emits at 4700 A arises from electron scattering and emission from the negative hydrogen ion. This ion consists of a hydrogen atom, with an additional electron tacked on. The binding energy of the extra electron is so small that visible light photons can easily detach it. Thus the negative ion (H^-) absorbs and emits visible light very efficiently in the photosphere and low chromosphere. The continuum intensity at 3640 A arises from a different atomic process, the emission of hydrogen atoms. Each intensity depends on temperature (T) and density (N) in a different way. In effect, these scientists had two unknowns (T,N) at each height in the chromosphere and they had two observed quantities (the light intensity at the two key wavelengths) with which to determine the unknowns.

Figure 4.1 illustrates their famous "plateau" model of temperature. The temperature rises from a minimum of about 4700°K just above the top of the photosphere to a plateau of 6000°K. The

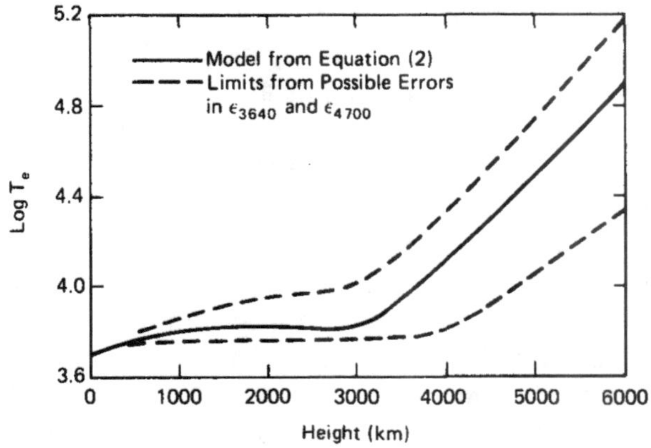

Fig. 4.1. The height variation of temperature in the chromosphere, determined from observations of the flash spectrum in 1952. (*Reprinted Courtesy of R. Grant Athay and* The Astophysical Journal, *published by the University of Chicago Press; © 1955 The American Astronomical Society.*)

temperature then rises abruptly at 3000 km, presumably to reach coronal values of one or two million degrees.

This model made a lot of sense. Hydrogen gas, which is the main constituent of the chromosphere, ionizes (i.e. splits up into protons and electrons) only when its temperature reaches about 6000°K. Moreover, hydrogen radiates energy very efficiently when it is partially ionized (when, say, 50% of all hydrogen atoms are neutral and the rest ionized). Thus the existence of a plateau suggested that the chromospheric temperature rises so that the gas can radiate the energy that is deposited by shock waves or by any other nonthermal source.

With this excellent start, the consortium of three institutions prepared for the eclipse of 1958 in Puka Puka, a tiny atoll in the Cook Islands. The members of the expedition rebuilt the eclipse spectrographs, providing three cameras to extend the spectrum into the red. As it happened, they couldn't complete the instrument before the deadline to ship it, so they were forced to carry out the final alignments and the fussy adjustments to the cameras at the eclipse site. They worked literally day and night, in shifts, for two weeks before the eclipse. On eclipse day, they were all exhausted and barely cared whether the experiment would succeed. Perhaps this cavalier attitude offended the South Sea gods. In any event, the eclipse was clouded out. They packed their huge cases and returned home.

The next opportunity appeared in 1959, only a year later. This time they chose a site on Fuerte Ventura, in the Canary Islands. More travail, more last minute adjustments at a feverish pace. Once again, clouded out! Frank Orrall, a member of the observation team, has said that they were actually relieved at the clouds, for fear their equipment might not work!

They had learned their lesson. Never again would they take such complex equipment to an eclipse and expect to make it work on the spot. They completely revamped the spectrograph cameras, reducing them to two: an "ultraviolet" camera that covered 3100 A to 6700 A and an "infra-red" camera, that covered 3300 A to 9100 A. They tested and retested until they were sure the machine would work whether or not the eclipse was clear.

Once again our brave troupe set forth, this time for the eclipse of February 5, 1962 in Lae, New Guinea. Figure 4.2 shows them, relaxed and smiling on the beach after the eclipse. They are smiling because the sky cleared only minutes before totality and because their equipment recorded a superb set of spectra (see Figure 4.3).

Fig. 4.2. The joint expedition of the High Altitude Observatory, National Bureau of Standards and Sacramento Peak Observatory, at Lae, New Guinea, 1962. They are posing with members of the Kyoto University Expedition.

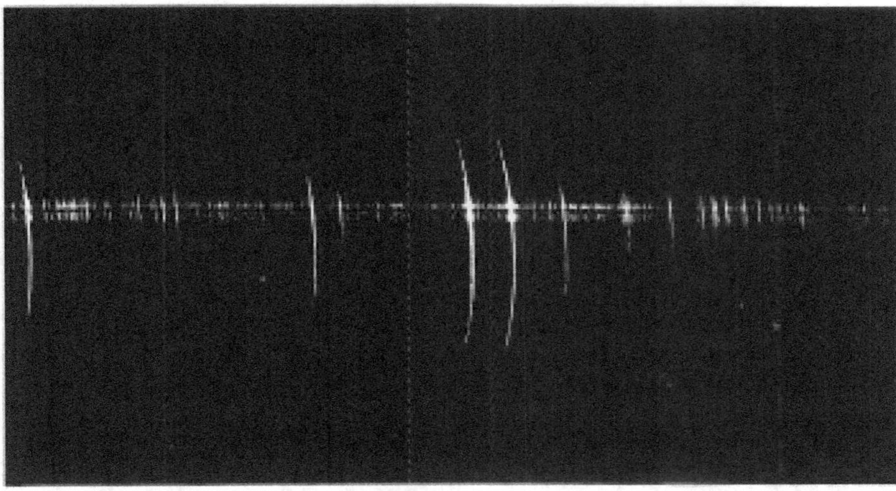

Fig. 4.3. A portion of the chromospheric flash spectrum, photographed during the 1962 eclipse. Wavelength decreases from left to right. Each crescent is an image of the chromosphere in the light of a particular spectrum line. The two long, strong lines in the center of the picture were emitted by singly-ionized calcium.

By 1962, the members of the joint expedition began to realize how fiendishly difficult it is to analyze the line spectrum of the chromosphere (as opposed to *continuous* spectrum), and to improve the empirical distributions of temperature and density. They had run into a whole forest of theoretical obstructions and complications. The observations were, at last, ahead of the theory. They decided to pause and digest their data.

Meanwhile, astronomers at the Tokyo Astronomical Observatory had realized how to improve on the design of the jumping-film spectrograph. As we pointed out earlier, the American version used no slit. As a result, each spectrum line on the film is, in effect, a monochromatic image of the chromospheric crescent. The width of the spectrum line is determined mainly by the amount of chromosphere showing above the Moon's limb and *not* by the intrinsic qualities of the gas, such as its temperature. By tilting the grating of the spectrograph to an unusually high angle, the Japanese were able to demagnify the width of the chromospheric image to a point where it is smaller than the intrinsic (i.e. temperature-broadened) width. Thus, the Japanese could record the intrinsic spectrum line profiles (i.e. intensity versus wavelength), not merely the total intensity of a line. This imaginative advance promised to give them a much more direct means of estimating chromospheric temperatures.

They first tried out their idea at the eclipse of October 12, 1958 in the Suwarrow Islands in the South Pacific, using two cameras to photograph the wavelength region from 3590 A to 6600 A. The Japanese were troubled by clouds during totality and third contact, but managed to capture a fine series of spectra at second contact. However, they did not analyze them, for some reason.

Not to be outdone by their colleagues at Tokyo, a team from Kwasan Observatory in Kyoto built their own jumping-film spectrograph and observed the flash spectrum at the February 5, 1962 eclipse in Lae, New Guinea—not far, in fact, from the American team. The Kyoto group concentrated on two spectral regions: 5876 A to 6563 A and 8100 A to 8600 A, but the latter was insufficiently exposed and the former contained strong lines only and no "easy" continuum.

The best Japanese flash spectra were secured at the Peruvian eclipse of November 12, 1966. E. Hiei and Z. Suemoto led a team from Tokyo Observatory. Once again, the Japanese had good luck in choosing a site, and their experiment went off without a hitch. They photographed the spectrum between 3580 A and 4180 A, at intervals of 380 km, up to a height of 50,000 km, above both limbs.

K. Tanaka and E. Hiei reduced and analyzed the blue continuum data in much the same way that Athay and Thomas had, taking into account improvements in the theory of ionization. Their temperature model, based primarily on the H^- continuum data, extends only to 1000 km. Limited to heights below 1000 km, the Tokyo model could not confirm the existence of the temperature plateau at 6000° to 7000° that Athay and Thomas had found.

Since everyone agreed that continuum observations were much easier to analyze than line spectra, the Japanese resolved to improve their data with new observations, and to try, in particular, to extend them to greater heights, where the chromosphere breaks up into spicules.

Spicules are thin (800 km diameter) jets of gas, that poke out of the top of a nearly homogenous layer that extends to perhaps 3000 km. Perhaps the most dramatic photograph ever made of spicules was made by R. B. Dunn (Figure 4.4) outside of eclipse with a narrow-

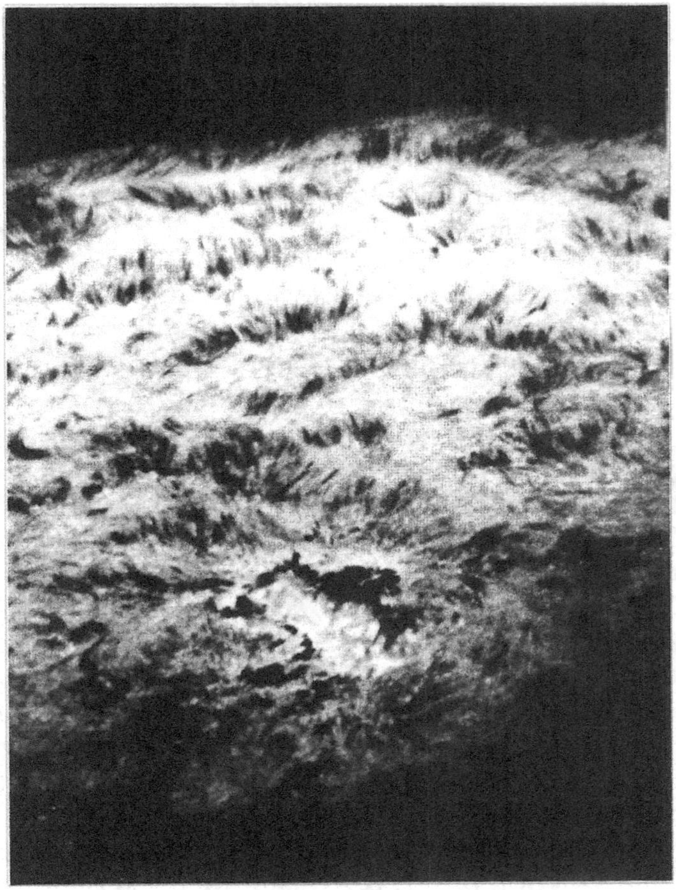

Fig. 4.4. Spicules near the limb of the Sun.

band, hydrogen line filter. As you can see, the needle-like spicules cluster along the boundaries of the supergranulation cells. They rise and fall with speeds of about 25 km/s. Above a height of about 5000 km, nearly all the chromospheric emission arises in the spicules. The interiors of the cells are virtually bare of chromospheric material and the hot corona practically touches the photosphere there with no intermediate layer.

The Tokyo Observatory sent a team, headed by M. Makita, to photograph the chromosphere at the eclipse of March 7, 1979. Makita located his equipment at Puerto Escondido, a small village on the Pacific coast of Mexico. His instrument could not be simpler. It consisted of a ceolostat (two flat mirrors that deflected sunlight into a horizontal telescope), an objective lens, a narrow-band filter and a Nikon motorized camera. The filter and film sensitivity restricted the band pass of the instrument to 200 A of continuum, centered on 6900 A. As usual, the Japanese had fine weather. Makita obtained a fine series of monochromatic images of the chromosphere that extend up to a height of 2500 km. Unfortunately, the "seeing" during the eclipse *was only fair and no details on the limb sharper than about 1700 km* were recorded. Most of the spicules that Makita was hunting were badly blurred on the photographs.

Nevertheless, Makita determined the height distribution of the continuum intensity and compared it with the predictions of existing temperature models. The best model was that of R. Noyes and W. Kalkoven (see Figure 4.5), which matched the observations beautifully to a height of 1600 km, but predicted too low intensities higher.

It is fair to say that Makita's main goal of measuring the continuum intensity in individual spicules, was foiled by the poor seeing at eclipse time. These data are so valuable, however, that another team, led by R. Kopp of the High Altitude Observatory, planned to repeat his experiment at the June 30, 1973 eclipse.

They set up a ten-inch Cassegrain telescope on the east shore of Lake Rudolph in northern Kenya, on a small peninsula named Koobi Fora. This site had two prime advantages. First, the waters of the lake were expected to stabilize the atmosphere during eclipse and so preserve the seeing. Second, the site lay only 4 km inside the northern limit of the path of totality. As a result, Kopp and company saw a *grazing* total eclipse of approximately 80 seconds' duration, which allowed them about four times the usual interval in which to photograph spicules.

The sky was clear during the eclipse but the seeing was "disappointingly poor." In fact, it deteriorated to 4 to 5 arcseconds (3000 km to 4000 km) during the eclipse—much too poor to resolve individual spicules. So much for the magical properties of lake water!

Despite their poor spatial resolution, the HAO photographs showed continuous emission in column-like structures (unresolved spicules, presumably) up to heights of 10 to 12000 km—much higher than normally expected. Kopp and his colleagues did not attempt to derive the temperature distribution (it would have been difficult with

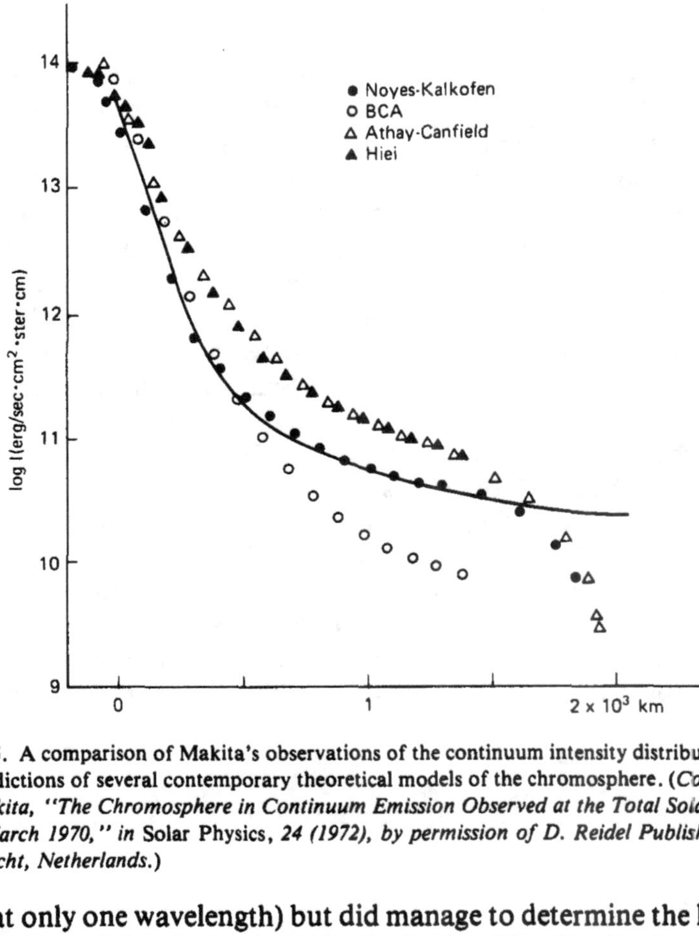

Fig. 4.5. A comparison of Makita's observations of the continuum intensity distribution with the predictions of several contemporary theoretical models of the chromosphere. (*Courtesy of M. Makita, "The Chromosphere in Continuum Emission Observed at the Total Solar Eclipse on 7 March 1970," in* Solar Physics, *24 (1972), by permission of D. Reidel Publishing Co., Dordrecht, Netherlands.*)

data at only one wavelength) but did manage to determine the height-distribution of density in unresolved spicules. The distribution agreed moderately well with other determinations.

At the same time Kopp was photographing the chromosphere, Soviet astronomers were taking jumping-film spectra. For some reason, they never analyzed their data. Thus, the HAO expedition was the last productive eclipse experiment on the visible light spectrum of the chromosphere. Astronomers have turned to other techniques since then: ultraviolet spectra from satellites such as Skylab and the Solar Maximum Mission, groundbased coronagraphs and narrow-band filters, and, as we shall see, eclipse measurements at millimeter wavelengths.

Limb-brightening

Before we consider recent millimeter-wave eclipse experiments, we need to understand some simple physics that they rely on. These experiments seek to determine the run of temperature versus height in the chromosphere by measuring the intensity of sunlight at one or more wavelengths in the continuous spectrum as the Moon covers or uncovers the Sun. The intensity of emerging light is determined by the temperature of the layer of gas that last emitted the light. Because the solar photosphere is relatively opaque, light can escape the Sun only from layers that lie at shallow depths—perhaps 100 km below the "surface." In effect, we "see" into the Sun only a short distance. As we look toward the limb, our line of sight slants into the spherical layers (see Figure 4.6) so that we see down to *shallower* depths near the limb than at the center of the disk.

If we look in visible light at the disk center and then at the extreme limb, we notice that the disk is *brighter* and *bluer* than the limb. Since a gas emits more light and bluer light at higher temperatures, and since we see deeper in the Sun at the disk center, we can immediately conclude that the *temperature decreases outward* in the photosphere.

But at some height near the top of the photosphere, the temperature

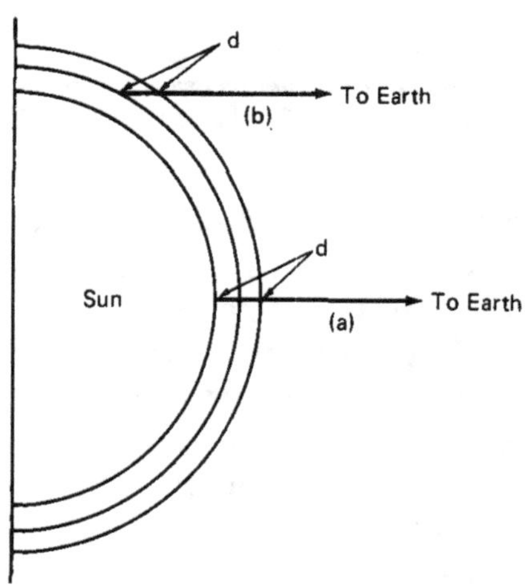

Fig. 4.6. The light that escapes from the center of the solar disk (*a*) originates within a depth *d* below the surface. Light from deeper layers is blocked by the overlying opaque gas and must diffuse outward slowly. At the limb, light emerges through the same distance *d*, but because of the photosphere curvature, originates in *shallower* layers.

reaches its minimum and begins to *rise,* in the outward direction, in the low chromosphere. What do we expect to see now as we look toward the limb? If you answer "brighter and bluer light," you would be correct—except for another complication: the density (and therefore the *opacity*) of the solar gas decreases at greater heights. At some height, no matter which wavelength we choose, we will see *completely through* the photosphere and low chromosphere. In fact it is for this reason (i.e. the decrease in density) that the Sun has such a sharp edge in visible light. But the edge is much less sharp at millimeter (and longer) wavelengths.

Figure 4.7 illustrates how the opacity of the solar gas changes with wavelength, around the temperature minimum. The gas is most transparent at 1.6 microns, and becomes more opaque at both shorter and longer wavelengths. In particular, its opacity increases steadily toward the millimeter wavelengths. The gas owes its opacity mainly to the absorbing power of the negative hydrogen ion which forms at the relatively cool temperature and relatively higher pressure of the photosphere and low chromosphere.

Now, let's combine the factors that determine the brightness of emergent light: the temperature and density along the line of sight and the intrinsic, wavelength-dependent absorbing power of the gas. Im-

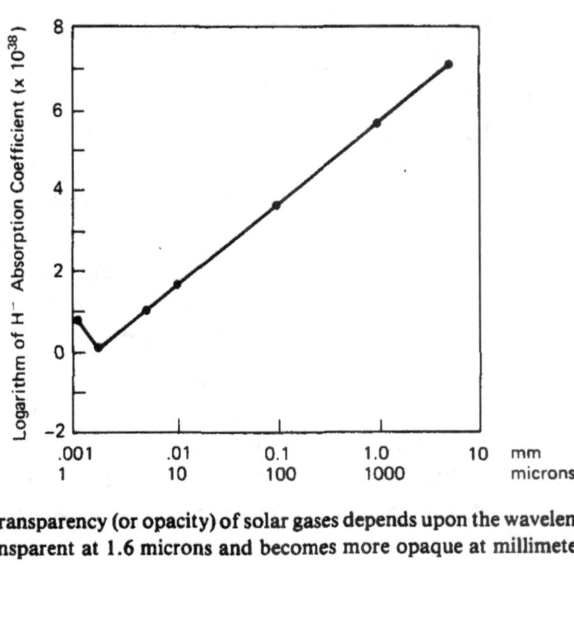

Fig. 4.7. The transparency (or opacity) of solar gases depends upon the wavelength of light. The gas is most transparent at 1.6 microns and becomes more opaque at millimeter wavelengths.

agine scanning outward, across the extreme limb of the Sun with a detector that is sensitive to millimeter waves. At first, it will record *increasing* brightness toward the limb because the temperature rises in the low chromosphere. But at some point, the gas along the line of sight will become transparent, because the gas density decreases outward, and the detector will sense *less* light. In fact, the brightness will drop rapidly toward greater heights. However, the intrinsic gas opacity is higher at longer wavelengths (see Figure 4.7), so the brightness drop will occur at greater heights for the longer wavelengths. Thus, if our physics and our picture of the temperature distribution in the chromosphere are correct, we expect to see a *bright ring* at millimeter wavelengths around the optical limb. The intensity profile of this ring should contain detailed information on the radial temperature distribution of the chromospheric gas.

The Chromosphere at Millimeter Wavelengths

Radio astronomers realized, from the foregoing reasoning, that they might be able to detect limb-brightening at millimeter wavelengths during an eclipse. In this region of the spectrum, few (if any) lines appear. The spectrum is continuous and as a result, much easier to interpret than the lines in the visible light flash spectrum. Their method is quite simple, in principle. They just measure the total flux of the exposed limb continuously, as the Moon covers or uncovers the Sun. When they correct their observations for their antenna pattern and for the known rate of change of area of the crescent, they find a profile of intensity (or brightness) versus radius for the Sun. If a "spike" of brightness exists, it should show up in the profile. The smooth motion of the Moon guarantees that their data will have a spatial resolution of an arc-second or better.

In practice of course, the experiment is more difficult. A ground-based eclipse experiment must contend with the low transmission of the Earth's atmosphere at millimeter wavelengths, caused by water vapor. Since most of the chromospheric light is absorbed, the remaining signal tends to be noisy. Moreover, the Moon and the atmosphere itself contribute to the observed signal. In order to isolate the chromospheric contribution, astronomers arrange their equipment to look alternately at the chromospheric crescent and at the sky away from the crescent. Their electronic equipment automatically subtracts the latter from the former signal, to yield only the chromospheric contribution.

Exotic detectors are needed to sense millimeter-wave solar radiation. Some of these are photoconductors, like indium arsenide, whose electrical conductivity changes according to the amount of incident light. Similarly, the filters and monochromators used to isolate narrow spectrum bands are quite special. At millimeter wavelengths, an experimenter can choose either a conventional optical telescope or a radio antenna "dish" to image the Sun on his detector. Either way, the telescope will accept light from an angle on the sky that is much larger than the thickness of the chromosphere, and suitable corrections must be made for this. Ultimately, of course, the Moon's motion provides high spatial resolution in the chromosphere.

The first successful experiment of this type was made by R. Coates of the Naval Research Labs at the total eclipse of June 20, 1954 in Sweden. He found a slight brightening within an arc-minute of the limb at a wavelength of 8.6 mm. Coates was followed by J. Hagen and his students from Penn State University, who observed the March 7, 1970 eclipse at 3.2 mm. They derived a curious *double-spiked* brightness profile from their data, that was probably due to noisy data, at least in part. P. Swanson and J. Hagen tried again with better digital recording equipment at the June 30, 1973 eclipse, just south of Atar, Mauritania. Despite equipment troubles during the eclipse, they detected a single strong spike, just inside the optical limb.

F. Shimabukuro, from the Aerospace Corporation in California, observed the same eclipse at 3.3 mm from Lake Rudolph, in Kenya. He also found a spike, 20% brighter than the disk, a half-arcminute inside the optical limb. He compared this result with predictions from the Harvard Smithsonian Reference Atmosphere, a notable contemporary model of temperature and density in the upper layers of the Sun. The model predicted too much emission inside the limb and also predicted that the peak of the spike would lie 2.5 arcseconds outside the optical limb, in conflict with the observations. Clearly, the model needed modifications.

Figure 4.7 shows that the chromosphere becomes more transparent at wavelengths shorter than 3 mm. Would it be possible to detect limb-brightening below 3 mm? Astronomers realized that, in order to accurately map the temperature in the low chromosphere, they needed to find the shortest wavelength at which limb-brightening (i. e., a limb "spike") exists.

J. E. Beckman, a physicist from Queen Mary College, London, decided to try at the eclipse of June 30, 1973. He built an instrument to scan the spectrum rapidly between 0.4 and 1.2 mm. It consisted of a Michelson interferometer[1] instead of a spectrograph, an indium arsenide detector, a 12 cm diameter telescope, and two guiding mirrors. This equipment could produce a complete spectrum once a second.

Beckman flew his equipment aboard the Concord 001, the superb British-French supersonic jet, at an altitude of 57,000 feet. The jet streaked down the path of totality, keeping pace with the Moon's shadow. Beckman enjoyed a 65 minute totality. The air above the jet was practically free of water vapor, which usually blocks out submillimeter radio waves at ground level. Conditions for the experiment were near-perfect and although Beckman encountered equipment trouble (inevitably *during* the eclipse!) he managed to obtain some very precise data. Figure 4.8 shows the flux he measured at three wavelengths, at different positions of the lunar limb. Note the small error bars on his points and the close spacing (1.6 arcseconds). Figure 4.9 illustrates the limb spike that Beckman derived from his data. The spike sits practically *at* the limb, extending only 4 arcseconds outside and 5 arcseconds inside. The peak brightness (i. e. temperature) is the same (10,000°K) at all three wavelengths.

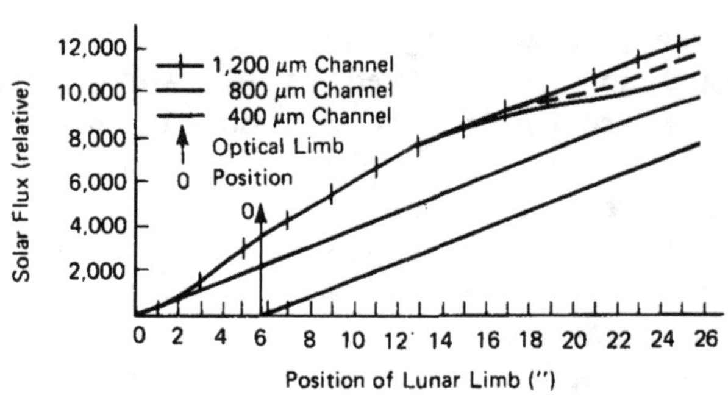

Fig. 4.8. Beckman's 1973 observations of the radial brightness distribution of the chromosphere, at submillimeter wavelengths. (*Reprinted by permission from* Nature, *Vol. 254, March 6, 1975 p. 39, Copyright © 1975—Macmillan Journals Limited, London.*)

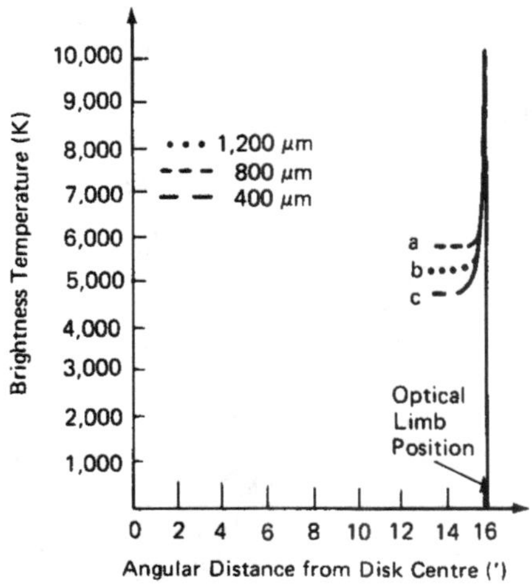

Fig. 4.9. The "spike" of intensity at the limb of the Sun, as determined by Beckman during the 1973 eclipse. (*Reprinted by permission from* Nature, *Vol. 254, March 6, 1975, p. 39, Copyright © 1975—Macmillan Journals Limited, London.*)

Beckman concluded from this last characteristic that the limb spike arises in chromospheric spicules. Contemporary models, based on optical spectra, implied that individual spicules should be opaque at sub-millimeter wavelengths. Although spicules cover less than 1% of the solar disk, they overlap at the limb. In effect, one sees through a forest of spicules in front of and behind the limb. The models suggested that this fringe of spicules could produce a spike, with a temperature near 10,000°K, at sub-millimeter wavelengths.

Beckman's neat results remained unchallenged for only six years. T. A. Clark and Rita T. Boreiko, physicists at the University of Calgary decided to check Beckman's important findings at the shortest (and most critical) wavelength of 0.4 mm. They flew their equipment, a "simple" photometer, aboard the NASA Lear Jet Observatory during the eclipse of February 29, 1979 in North Dakota and southwestern Manitoba. Once again the jet's speed and altitude created excellent observing conditions. Unfortunately, aircraft motions and steering problems reduced the total effective observing time to 7 minutes, centered on second contact. In Figure 4.10, we compare their measurements of flux versus time, with predictions from the Harvard Smithsonian model, from a perfectly flat temperature model, and from another temperature model with a chromospheric rise. Clark

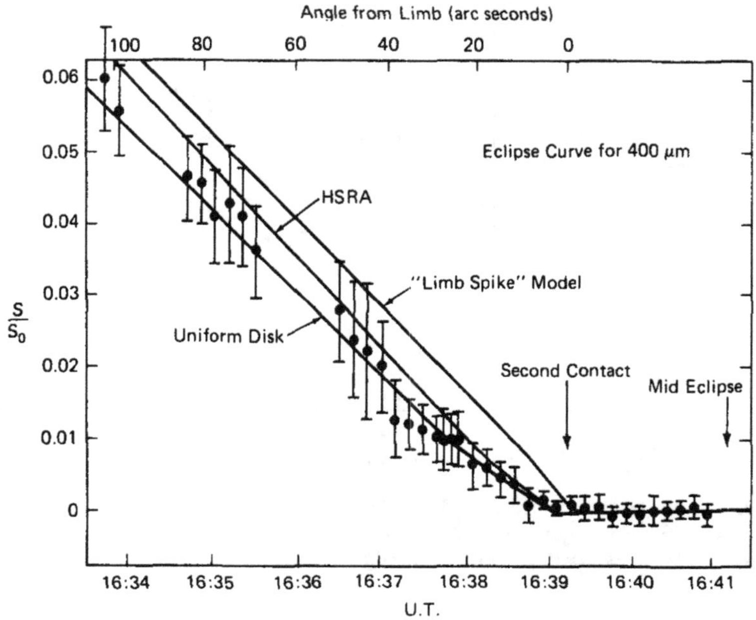

Fig. 4.10. Observations of the radial brightness distribution of the chromosphere obtained by Clark and Boreiko at the 1979 eclipse. Several theoretical brightness models are shown for comparison. *(Courtesy of T. Alan Clark, "Airborne Total Eclipse Observation of the Extreme Solar Limb at 400 μm," in* Solar Physics, *76 (1982), by permission of D. Reidel Publishing Co., Dordrecht, Netherlands.)*

and Boreiko concluded that a flat temperature model (i. e. a solar disk of uniform brightness, with *no* limb-brightening), fits their data best! Their results flatly contradicted those of Beckman. According to Clark and Boreiko, spicules are *not* observable inside the limb, nor do they contribute to any limb-brightening at 0.4 mm.

As of this writing, the contradictions between Beckman's results, and Clark and Boreiko's results, have not been resolved. A new experiment will have to be mounted at a future eclipse to settle the matter.

At wavelengths shorter than a few tenths of a millimeter, the low chromosphere becomes increasingly transparent (see Figure 4.7), and a limb-scan at shorter wavelengths will sample progressively deeper layers. At some wavelengths, in the vicinity of 0.1 millimeters or 100 microns, we expect to detect the presence of the temperature minimum zone, just above the photosphere. The temperature at the minimum, its width and the gradient above it, are important quantities that determine the amount of nonradiative energy that is deposited in this region.

An experimental team from the University of Hawaii (E. Becklin, C. Lindsey, J. Jefferies, F. Orrall) devised an experiment to investigate the temperature minimun zone during the eclipse of July 31, 1981. They too flew aboard a high-flying jet—the NASA Ames Kuiper Airborne Observatory—along the path of totality, which crossed the western Pacific. They observed the eclipse simultaneously in four wavelength bands: 30, 50, 100 and 200 microns. The 36-inch telescope aboard the jet looks out of the side of the plane, without an intervening window. The team was required therefore to wear oxygen masks and warm clothing for protection against the frigid air temperature.

The Hawaiian experiment was a success, but only the following few details of their results have been published as yet. No limb-brightening was observed at any of the four wavelengths. Instead, they found that the *diameter* of the Sun was progressively larger, the longer the wavelength. Thus, for example the limb at 200 microns lay 2.5 arcseconds above the limb at 30 microns. Perhaps the most surprising feature of their results concerns the variation of limb-*darkening* with wavelength. The brightness of the extreme limb at each wavelength decreases with increasing distance from the center of the Sun, but the decrease is *greater* for the *longer* wavelengths. For example, the intensity of the limb at 200 microns was about 10% less than at 30 microns. This behavior is opposite to that at visible wavelengths, where the amount of limb-darkening is larger for blue (short) light than for red (long) light.

The Hawaii data evidently sample the temperature minimum and the beginning of the chromospheric temperature rise. The team is struggling now to understand the observed increase in the solar radius with increasing wavelength. If this "limb-extension" is attributed to the chromospheric spicules (which we know extend thousands of kilometers beyond the visible limb), then the spicules must be opaque at 200 microns and transparent at 30 microns. These inferences from the data contradict current models of temperature in these low layers of the chromosphere. Moreover, models that do not take into account in spicular nature (e.g. the inhomogeneity) of the chromosphere predict limb-*brightening* at 200 microns, in strong contradiction to the observations.

A new era of airborne eclipse observations in the submillimeter range of the flash spectrum has evidently begun. We can look forward to more exciting results at future eclipses.

NOTES

1. The Michelson Interferometer

The interferometer is a device used for analyzing infra-red and millimeter wave spectra. Figure 4.11a is a sketch of its main elements.

Light from a source (S) divides at the inclined "beam-splitter" (BS). Half goes to mirror M_1, reflects and returns to a light-intensity detector (D)—e. g. a photocell. The other half reaches D after reflection from mirror M_2. The two beams "interfere" at D: they *reinforce* if the path-length BS to M_1 to D equals the path-length BS to M_2 to D and *cancel* if the path lengths differ by half a wavelength.

In practice, M_2 oscillates in position, changing the pathlength in one "arm" of the interferometer. As a result, the detector might record a signal as in Figure 4.11b.

The presence of the intensity peak indicates that the source strongly radiates a particular wavelength, that can be determined from the position of M_2. The other wiggly sections of the curve in Figure 4.11 indicate that many other interferring wavelengths

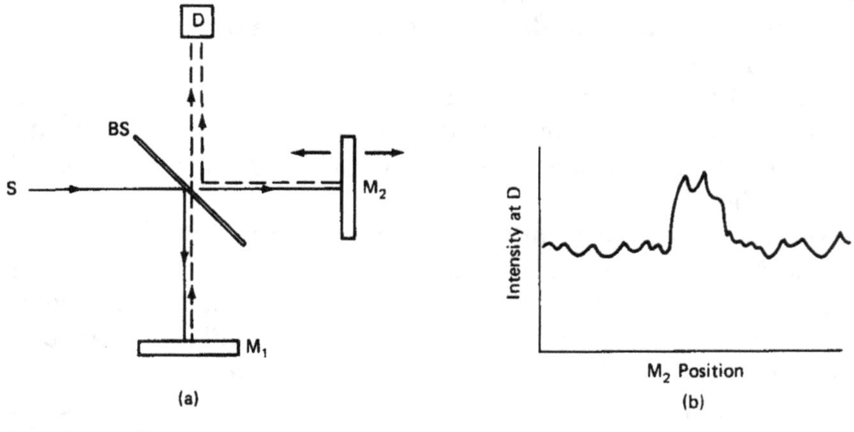

Fig. 4.11. (a) The principle of the Michelson interferometer. (See text for explanation.)

(b) When the mirror, M_2, oscillates, the signal at the detector, D, varies as shown. The intensity peak indicates the presence of a strong, monochromatic spectrum line in the light from the source, S.

are also present in the incident light. Mathematical analysis by Fourier techniques will yield the intensity at each wavelength (e. g. a real *spectrum*).

SUGGESTIONS FOR FURTHER READING

Books

Athay, R. G. *The Solar Chromosphere and Corona.* Reidel Publishing Company: 1976.

Athay, R. G. "The Chromosphere and Transition Zone," in *The Sun as a Star.* S. Jordan, editor. NASA-SP-450: 1981.

Noyes, R. *Our Sun—A Star.* Harvard University Press: 1982.

Periodicals

Withbroe, G. and Noyes, R. "Energy and Mass Balance in the Solar Chromosphere Corona," in *Annual Reviews Astronomy and Astrophysics* **15**, 363, 1977.

5
The Solar Corona

INTRODUCTION

The past fifteen years have been an incredibly productive period in developing our understanding of the physical nature of the solar corona. Total eclipses of the Sun have continued to offer astronomers outstanding opportunities for investigation. Non-eclipse observations from satellites, rockets, and ground-based coronagraphs have, however, played an increasingly important role, and together with eclipses have revealed the complicated behavior of the corona. The new observations have, as usual, raised as many questions as they have answered. Astronomers who study the corona are, in effect, unpacking a set of nested Chinese boxes, each one smaller and more intricate than the preceding one, and more difficult to reach.

By the mid-1940s, astronomers had acquired most of the basic ideas on the corona that they accept today. The corona was pictured as an enormous extension of the visible atmosphere of the Sun. The solution of the "coronium" puzzle proved that the corona is very hot (two million degrees or more). It is an extremely thin gas, with a pressure a million times smaller than the Earth's atmosphere. Because the corona is so hot, most of its hydrogen is almost completely ionized into protons and electrons, and the traces of heavier elements it contains, such as helium, oxygen, neon, carbon, and all the familiar terrestrial metals, are also very strongly ionized.

Coronal structures seem to conform to the shape of the coronal magnetic field, although the proof of this statement, as we shall see, has only been advanced recently. The earliest, and most obvious connection between coronal forms and coronal magnetic fields was the observed variation of coronal structures throughout the 11-year sunspot cycle. As you will remember, Schwabe discovered this cycle in 1843, and Hale showed that sunspots are basically regions of strong magnetic fields.

The corona meets the chromosphere in an intermediate layer, the "transition region." Here the temperature rises steeply outward. Compared with the quiescent corona and photosphere, the transition region churns furiously. The "normal" physics of a hot gas breaks down in this region (as in the chromosphere), where every atomic process indicates a different temperature.

The Sun and its planets inhabit a dust cloud of tiny ice particles that are probably the remnants of the original solar nebula. At great distances from the Sun, these particles scatter sunlight to produce the familiar zodiacal light (see Figure 6.1, page 107). Close to the Sun, sunlight scattered by these particles contributes to the brightness of the corona during a total eclipse. The intense heat of the Sun evaporates all the particles within two or three solar radii of the Sun's surface.

Two major developments occurred during the 1960s and 1970s. The German astrophysicist Ludwig Biermann was struck by the fact that the gaseous tails of comets that pass through the inner solar system point directly away from the Sun. When he looked into the matter, he found that the accepted explanation—that the pressure of radiation from the Sun forces the gas radially outward—was inadequate. He postulated instead a continuous flow of ionized gas from the Sun that drags out the cometary gas into a long tail. Fluctuations in the solar flow could account for observed fluctuations in comet tail brightness. The geophysicist J. Bartels had postulated a flow of solar material through interplanetary space as early as 1932 in order to account for geomagnetic storms that recur at intervals of the solar rotation period, 27 days.

Eugene Parker, at the University of Chicago, picked up on Biermann's idea. He was interested in the hydrodynamical properties of a hot, tenuous stellar envelope, like the solar corona. His theoretical studies showed that a hot corona must inevitably expand into the near vacuum of interplanetary space. The resulting flow could reach a supersonic speed of several hundred kilometers per second past the Earth. Parker's predictions were confirmed brilliantly by experiments aboard the spacecraft Mariner II in 1962. They showed not only that the flow exists but that it can concentrate into long-lived, fast "streams," with speeds up to 800 km/s. These streams were identified as the probable causes of recurrent geomagnetic storms.

The "solar wind," as the flow of ionized gas was labeled, became firmly established as the fundamental medium which carries mass, energy, and momentum away from the Sun and deposits them in the environment of the Earth.

For a decade following these developments, astronomers searched unsuccessfully for the places in the solar corona where the solar wind originates. Solar streamers were an obvious first choice, but they did not reappear in cadence with geomagnetic storms.

Radio observations of the Sun did not help much. Although radio astronomers were able to map the corona that lies directly above the disk of the Sun, they were hampered by insufficient spatial resolution and sensitivity (i.e., their telescopes were too small and too slow), by uncertainties in the physical processes that produce the radio radiation, and by the complicated paths that radio waves follow in the corona because of density inhomogeneities and magnetic fields.

The question remained open until the mid-1970s, when space observations provided the essential clue. A series of Orbiting Solar Observatories ("OSO's") launched by NASA in the early and mid-sixties began to give increasingly sharp pictures of the corona that overlies the visible disk. These satellites mapped the corona at ultraviolet wavelengths (shorter than 1000 A) or x-rays (shorter than 100 A). Space observations confirmed what had long been apparent to ground-based astronomers: the corona is an extremely inhomogeneous and time varying structure.

In 1972, R. Munro and G. Withbroe constructed maps of the extreme ultraviolet brightness of the corona, using data from the Harvard University Experiment aboard OSO-4. They detected regions of the corona that were abnormally faint, and called them "coronal holes." Eclipse observers had occasionally seen such holes at the limb of the Sun, but thought they were accidental gaps between bright regions. The OSO-4 maps proved the holes were genuine, persistent regions of low density in the corona. W. Neupert's experiment on OSO-7 showed that some coronal holes persist for many months, and rotate with the rest of the Sun's atmosphere. Neupert and his collaborator, V. Pizzo, demonstrated that the recurrence of geomagnetic storms on the Earth was synchronized with the rotation of the holes. This relation strongly suggested that the holes were the sources of the solar wind, since streams in the wind were probably the cause of the storms.

The talented group of experimenters at the American Science and Engineering Company developed x-ray telescopes to the point where they produced coronal images of extraordinary sharpness. Figure 5.1 (on page 79) is an example. In 1973, the A. S. & E. group launched an x-ray telescope on a rocket, and photographed a prominent coronal hole. This hole coincided with the appearance at Earth of a high-speed stream in the solar wind. A. Kreiger and his associates at A. S. & E. immediately concluded that the hole was indeed the source of the stream, which in turn produced a magnetic storm on earth.

In 1973, NASA launched Skylab, a manned space observatory equipped with sophisticated astronomical instruments for a continuous study of the corona during a nine-month period. Skylab in-

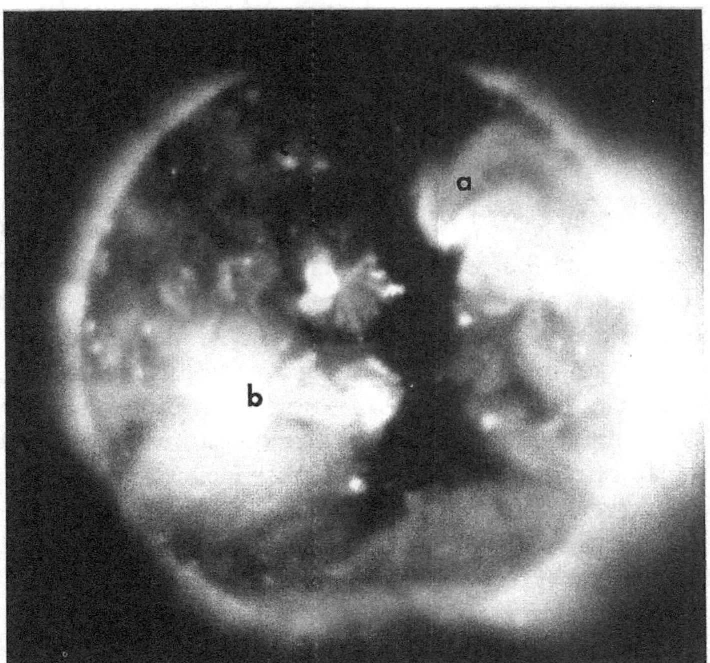

Fig. 5.1 The x-ray corona, photographed from Skylab. The large, dark lane (c) is a coronal hole. The other features are (*a*) coronal arches, (*b*) an active region. (*Courtesy of the Solar Physics Group, American Science and Engineering, Inc., Cambridge, Massachusetts 02139.*)

struments recorded the development and evolution of all the coronal forms, including streamers and holes. A detailed description of the marvelous results that came out of the analysis of Skylab's observations would take us too far afield. (See the suggestions for further reading at the end of this chapter.) Needless to say, the study of the corona was tremendously advanced.

The Orbiting Solar Observatories and their successor, Skylab, raised numerous questions which solar astronomers are presently working to answer. Among these are:

1. Why does the corona take the forms we see, and why do they change in phase with the sunspot cycle?
2. How is the corona heated?
3. How do coronal structures differ in temperature and density? In particular, how do holes and streamers differ?
4. How does material in the corona circulate? Does the corona rotate differently than the underlying photosphere? If so, why?

Let us see what eclipses and coronagraph observations, supplemented by the superb series of space observations during the past ten years, have to say on these questions.

CORONAL FORMS AND THEIR ORIGINS

In Figures 5.1 and 5.2, we have labeled the common characteristic forms of the solar corona as seen at the limb during a total eclipse and over the solar disk, as photographed at x-ray wavelengths.

We can divide the forms into categories according to size. The largest size includes the streamers and holes. Medium size features include active regions and coronal condensations which radiate brilliantly at x-ray wavelengths. Solar prominences, which are really coronal structures, differ in size according to their age and type. Active region prominences are typically small, with one end buried in a sunspot region. The large quiescent prominences are narrow, dark ribbons of relatively cool gas that wind through the inner corona for many hundreds of thousands of kilometers. As Figure 5.1 shows, the corona far from active regions consists of a tangle of loops, with varying size and brightness. Near the poles of the Sun, one can pick out very fine extended rays, called plumes.

Fig. 5.2 The white light corona, photographed at the June 30, 1973 eclipse. The identified features are (*a*) a streamer, (*b*) polar plumes, (*c*) a prominence at the limb. (*Courtesy of the High Altitude Observatory, National Center for Atmosphere Research, Boulder, Colorado—sponsored by the National Science Foundation.*)

What causes all this variety of form? Although there are several possible explanations, astronomers now agree that magnetic fields are mainly responsible for the shapes we see. Their conclusion rests on two strong arguments.

First, the shape of the corona at sunspot minimum (see Figure 5.3) resembles the magnetic field of a bar magnet or "magnetic dipole." Of course, no one seriously believes there is a bar magnet buried deep within the Sun, but the Sun obviously generates magnetic fields. George Ellery Hale discovered in 1908 that sunspots contain strong magnetic fields. Recent studies have shown that the rest of the Sun, far from spots, is covered with a lacy pattern of strong magnetic fields that concentrate into tiny clumps. Such fields will tend to rise and spread out into the corona. Thus, it is entirely *plausible* that magnetic fields reside in the corona.

The second reason why scientists believe that magnetic fields shape the corona relates to a property of an ionized gas, or "plasma." Such

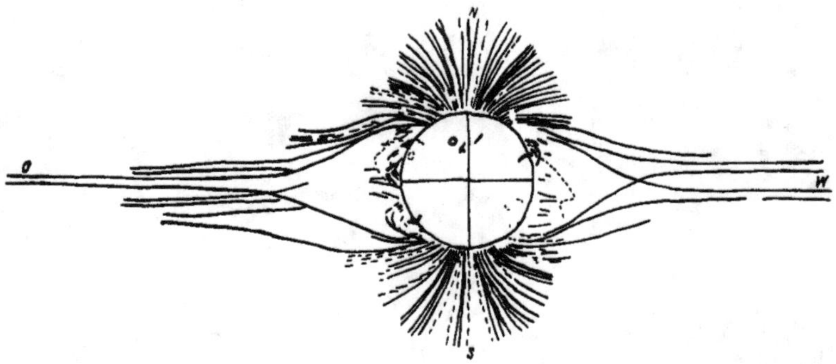

Fig. 5.3 A drawing of the corona, made from photographs taken at the June 30, 1954 eclipse. The corona was almost perfectly symmetrical at this minimum of the sunspot cycle. (*from S. K. Vsekhsvjatsky, "The Structure of the Solar Corona and the Corpuscular Streams," in proc. of IAU symposium no. 16,* The Solar Corona, *ed. by J. W. Evans, 1963, by permission of the International Astronomical Union.*)

a plasma cannot cross through magnetic lines of force without great difficulty. As soon as an ion or electron in the plasma attempts to cross, its charge interacts with the field to produce a force, the "Lorentz force," that bends the particle from its original path into a spiraling motion around the magnetic field line. Every attempt of the charge to pass through such a field results in a helical motion about the field, so the charges are trapped on the lines of force. They can move *along* the lines easily, but not through them. If the gas motions are sufficiently vigorous, they will drag the magnetic field, but if the magnetic field is sufficiently rigid, it traps and shapes the gas. The latter situation *must* prevail in the corona.

Now, while we can map the distribution of the magnetic field over the photosphere of the Sun with relative ease, coronal magnetic fields are extremely difficult to measure because coronal light is so faint. As a result, we have only indirect information about coronal magnetic fields, mainly from solar radio observations.

In 1964, H. U. Schmidt, at the Max Planck Institute for Astrophysics, suggested a way out of this impasse. He proposed a novel method for *calculating* coronal magnetic fields from measured photospheric magnetic fields. Schmidt assumed that electrical currents flow in the photosphere but not in the corona. The coronal field, then, is a simple extension of the field that breaks through the surface of the photosphere. Schmidt worked out the mathematical machinery required to extrapolate magnetic field lines that are rooted in the photosphere and that stretch into the corona.

G. Newkirk and M. Altschuler, scientists at the High Altitude Observatory, in Boulder, Colorado, extended Schmidt's work to apply to a sphere like the Sun rather than a plane. They then resolved to test the idea that coronal forms follow coronal magnetic field lines. They proposed to calculate coronal fields using Schmidt's technique, and to compare them with the coronal forms they could photograph in white light during a total eclipse of the Sun.

They obtained a beautiful photograph at the total eclipse of November 12, 1966. Using measurements of the photospheric magnetic field made at the Kitt Peak National Observatory throughout an entire solar rotation preceeding the eclipse, Newkirk and Altschuler calculated magnetic lines of force of the corona, and compared them with the structures that they had photographed. The two pictures are shown superposed in Figure 5.4. As you can see, the

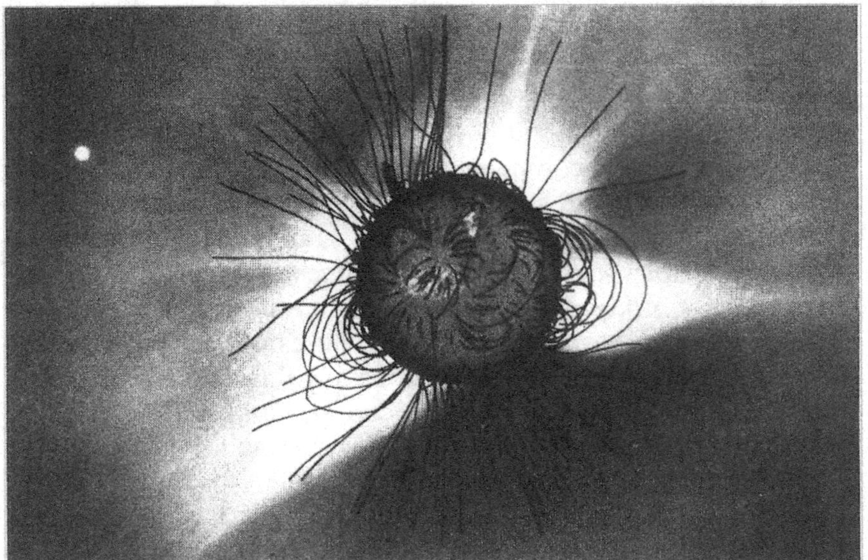

Fig. 5.4 Coronal magnetic fields, calculated for November 12, 1966, from measurements of photospheric fields taken over the entire Sun. The figure compares predicted coronal fields with white light features photographed at the eclipse. (*from* Structure and Development of Solar Active Regions, *proc. of IAU Symposium No. 35, Budapest, Hungary, September, 1967—by permission of the International Astronomical Union, and D. Reidel Publ. Co.*)

experiment was reasonably successful. Calculated field lines overlay the helmet streamers at the south-east and south-west limbs. "Open" field lines that stretch well out into interplanetary space fan out from the large region centered on the south pole. In the eclipse photo, you can see that these lines coincide reasonably well with the fine plumes around the pole. The overall direction of other calculated field lines agrees moderately well with the photographed structures, but many other features, even the large-scale high-latitude streamer in the northwest that shrinks into a narrow ray, do not show up in the magnetic field map. Nevertheless, you can conclude with some confidence that coronal magnetic fields create coronal structure, and that if we had a better way of calculating or measuring coronal fields, no doubt the correspondence would look better.

Newkirk and his associates continued to apply this method, with refinements, to coronal photographs they took at the 1970, 1973, 1980, and 1981 eclipses. Although they improved the detail with which they could calculate magnetic fields, the general correspondence between calculated fields and photographed forms did not improve appreciably. They were forced to conclude that, contrary to their basic assumption, electrical fields *do* flow in the corona, and therefore the magnetic field is more complex than they supposed.

Nevertheless, this method has proved extremely useful in guiding further research. The corona is seen to divide naturally into regions where the magnetic field lines loop through the corona and turn back to the Sun at low heights ("closed field"), and other regions where magnetic field lines stretch far out into space ("open field"). Closed-field regions coincide with solar active regions near sunspots or with the large bi-polar magnetic regions that develop from sunspots and which eventually extend into the corona as helmet streamers. Open-field regions invariably coincide with coronal holes, as J. Harvey, R. Levine, and N. Sheeley demonstrated with Skylab observations.

If electrical currents flow in the corona, as Newkirk and his colleagues demonstrated, the calculation of coronal magnetic fields become very difficult. Consequently, they tried to develop a method of measuring the actual fields, rather than calculating them. Their method depends upon the influence of magnetic fields on coronal light.

Astronomers had known for a long time that coronal emission lines, photographed at eclipse, are linearly polarized. The polarization arises because a coronal ion scatters atmospheric light (at the wavelength of its spectrum emission line) in *all* directions, while it *receives* light mainly from *one* direction, i.e. from the photosphere below it. Whenever a preferred direction prevails in the propagation of light through a medium, the light becomes partially polarized.

In 1977, L. House showed that a magnetic field in the corona will tend to *depolarize* the light that is scattered in a coronal emission line. The reasoning runs as follows. Ions in the hot corona tend to move in straight lines at high speeds. If a coronal ion encounters a magnetic field line, however, its trajectory is bent from a straight line into a spiraling helix, centered on the field line. Now imagine that such an ion absorbs a photospheric photon in one of its characteristic spectrum lines. Before it has a chance to re-radiate or scatter this photon, its helical motion "scrambles" the incident direction and counteracts the effect of the "preferred" direction from below. So, in effect, the ion loses any memory of the incident direction. The light it then radiates toward an observer will be partially *depolarized*. This process is called the Hanle effect.

House showed how to determine the direction and strength of a coronal magnetic field from the Hanle effect. In principle, the method yields information on the magnetic field only in the plane of the sky, and not in three dimensions. To reconstruct the full field, daily coronal observations would be needed through a full rotation of the Sun.

Quantum mechanics tells us which coronal emission lines polarize most strongly and, therefore, are the best candidates for this technique. The infrared line 10747 A of Fe XIII is one of the best. Accordingly, J. Eddy and his colleagues at the High Altitude Observatory measured the direction and degree of polarization of 10747 A at the total solar eclipses of 1965, 1966, and 1970. As expected, they found the degree of polarization fell to zero at the solar limb, but rose to more than 80% at a height of .6 solar radii above the limb. Eddy compared his observations with polarization vectors calculated by House. House, in turn, used a Newkirk-Altschuler model of calculated coronal magnetic fields. Given the uncertainties in the observations and the flaws in the coronal field model, the two results agreed fairly well.

As we mentioned earlier, a single snapshot at an eclipse is insufficient to reconstruct the three-dimensional magnetic field of the corona. Instead, one needs observations continuously over at least one solar rotation. Newkirk and his colleagues, therefore, set out to build a new type of coronagraph, the coronal emission line polarimeter, that would build on House's theory and Eddy's observations and enable them to measure the polarization of the coronal emission line 10747 A daily. This instrument was built and then operated at Sacramento Peak Observatory for 7 years, beginning in 1973. Figure 5.5 shows such polarization measurements in the vicinity of a helmet streamer. The directions of the vectors and the white-light structure conform convincingly.

The polarimeter provides an indirect measure of the coronal magnetic field, but the instrument is limited in two ways. First, each

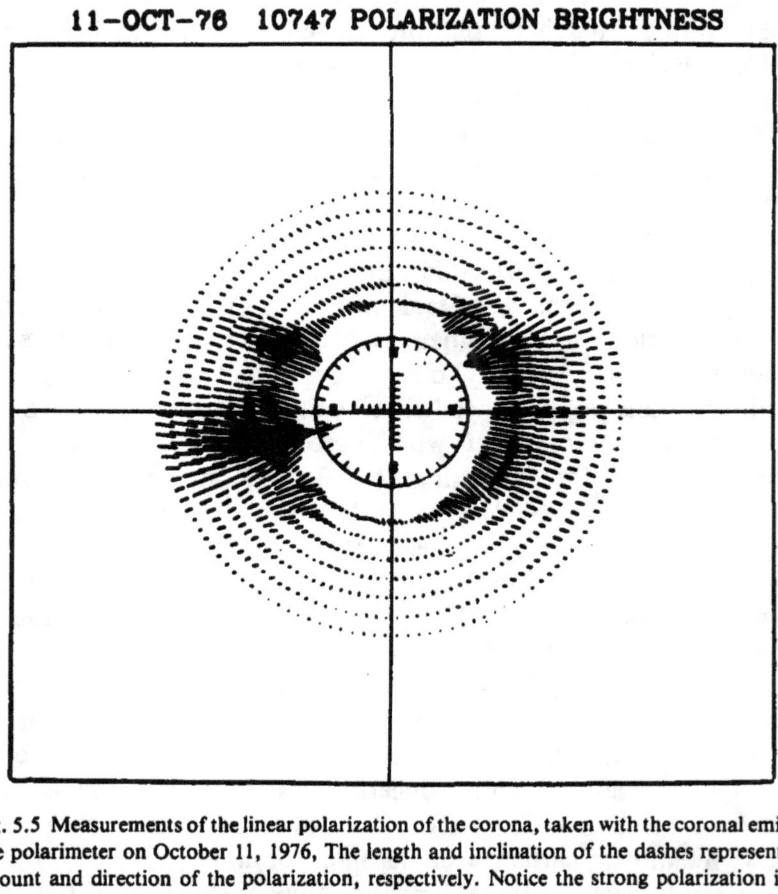

Fig. 5.5 Measurements of the linear polarization of the corona, taken with the coronal emission line polarimeter on October 11, 1976, The length and inclination of the dashes represents the amount and direction of the polarization, respectively. Notice the strong polarization in the streamers on the east (left) limb. *(Office of Rights and Permissions, Society of Photo-Optical Instrumentation Engineers, SPIE—The International Society for Optical Engineering, P.O. Box 10, Bellingham, Washington 98227, USA: Phone 206/676—3290, Pacific Time.)*

measurement is averaged over a rather large area on the corona, in order to collect sufficient light. Thus, many of the smaller details are lost. Secondly, a full rotation of the Sun must be observed to reconstruct the three-dimensional coronal magnetic field, and during a rotation, these fields probably change. Thus, the polarimeter yields, at best, time-averaged and spatially-averaged information. Nevertheless, the polarimeter is the best hope astronomers have at the moment of checking magnetic field calculations.

Newkirk and his colleagues found that the simplest kind of coronal magnetic field, i.e., one produced by electrical currents that flow only in the photosphere, does not match the coronal forms they photographed during total eclipses. In order to improve the match, one must somehow allow for currents that do flow in the corona and that modify the coronal field. Several attempts have been made in recent years to do this.

In one approach, the currents are postulated to flow only in flat sheets that separate large, current-free volumes. In another approach, the current is postulated to flow along the magnetic field lines. In both cases, the electrical current and the coronal magnetic field interact to produce a so-called "Lorentz" force on the conducting plasma. This new magnetic force, in combination with pressure and gravity, determines the distribution of coronal material and its circulation. As you can imagine, calculations like these, carried out for a realistic three dimensional corona become very complicated and very expensive. They have not come into general use.

THE TEMPERATURE OF THE CORONA

Once you recognize how much structure the corona contains (see Figure 5.1.), you can appreciate that no unique coronal temperature exists. Each loop and plume, each streamer and hole must have its own temperature distribution, which varies in all three dimensions. Faced with such complexities, the astronomer can only idealize the real situation and try to measure and model "typical" structures.

Even this goal is a fairly modern one. Until 10 years ago, astronomers were content with determining the average height distribution of temperature. Their change of outlook was motivated, of course, by increasingly better pictures of the corona at all wavelengths, and especially at x-ray wavelengths.

How does one determine the temperature of a coronal structure? There are several methods. The most generally useful depends upon the ionization characteristics of different chemical elements. At any fixed temperature, the atoms of an element are distributed among several stages of ionization, i.e. its atoms have lost one or more electrons. An element such as iron, for example, can exist in the corona as atoms with 9 electrons missing (Fe X), 10 electrons missing (Fe XI), and so on. If we could raise the temperature of a sample of coronal gas, the concentration of Fe X ions would first increase rapidly, reach a peak, and then decline as more and more of the available iron turned into Fe XI (see Figure 1.6 on page 22). Each of these ions radiates its own characteristic spectrum which we can identify, either experimentally in the laboratory or theoretically. When an astronomer observes the characteristic spectrum of, say, Fe X in some portion of the corona, he can be fairly sure that the temperature there approximates the temperature at which most iron atoms have lost 9 electrons, or about 1 million degrees Kelvin. If, on the other hand, the charateristic spectrum of Fe XV appears, the temperature must be closer to 2.5 million degrees Kelvin.

If, as usual, the spectra of several ions appear simultaneously, only two explanations are possible. Either the temperature lies at some intermediate value that permits several ions to exist in reasonable abundance simultaneously (see Figure 1.6) or the particular coronal structure being observed has a wide range of temperatures. This basic method of using characteristic spectra of elements in different stages of ionization has been applied throughout the entire electromagnetic spectrum, from gamma rays to the infrared.

The second way to estimate coronal temperature is to estimate electron density first. Free electrons in the corona scatter photospheric light. A photograph of the corona, when properly calibrated and interpreted, is a map of the distribution of coronal electrons. By measuring and analyzing such photographs, astronomers are able to infer the distribution of electron density, and therefore of gas density, in large or small structures. Next, the astronomer assumes that the gas in the structure is at rest or at least moving very slowly. In that case, the pressure in each layer of gas supports the weight of the layers above it. Such a situation is called "hydrostatic equilibrium." The density in an isothermal layer of gas decreases with increasing altitude; the density falls by a factor of 2.71 over a "scale height," H, where H

$= (k/mg)T$. Note that the scale height depends upon the temperature, T. Measure the rate at which density decreases with height, and you can determine a temperature.

The ions in a hot gas dance about at high speed; indeed, their speed is directly proportional to the square root of the gas temperature. If you can determine the average speed of an ion with a particular atomic weight, you can determine the gas temperature. This relation leads directly to a third method of estimating temperature.

An ion completely at rest would radiate spectral lines that are exceedingly narrow in wavelength. If the ion moved toward you, the wavelength of that line would shift to shorter wavelengths, i.e., it would be blue-shifted. Conversely, if the ion moved away the line would be red-shifted. This phenomenon is known as the Doppler effect. Along any line of sight through the corona, ions of any specific element will be moving at a variety of speeds, both away from and toward the observer. As a result, the spectral lines they emit will combine into a *broadened* emission line. The breadth of the line indicates the average speed or the kinetic temperature of the ions.

Once the coronium puzzle had been solved, and astronomers realized that they were dealing with a one or two million-degree corona, they readily accepted the temperatures determined from the observed scale height of coronal electrons. M. Waldemeier, the Swiss astronomer, has been especially diligent in photographing every solar eclipse since the early 1930s. From his photographs he extracted the density distribution, the shape, and a hydrostatic equilibrium temperature for the corona. Such eclipse results confirm the high temperature of the corona, and show a tendency, toward sunspot minimum, for the temperature at the poles to decrease, but the results give no information about the height variation of temperature.

The eclipse of 1952 yielded an extraordinary new result on the coronal temperature distribution. A joint expedition of the High Altitude Observatory and Sacramento Peak Observatory observed that eclipse in Khartoum, in the Sudan. A series of short photographic exposures was made of the spectrum of the chromosphere and corona, between second and third contacts. The spectrograph that these astronomers used was slitless, so a monochromatic image of the corona appeared on the film at the wavelength of each strong coronal emission line. As the moon moved it covered or uncovered the inner corona. When the team scientists studied their coronal images in detail, taking into account the motion of the moon, they realized that the corona begins at a height of no more than 7000 km above the limb of the Sun. In other words, the Sun's temperature rises abruptly by a million degrees or more between the photosphere and the base of the corona, in a distance of only 7000 km. Nobody could have guessed that the Sun's temperature rises so steeply to the base of the corona. Unfortunately, the emission line photographs did not extend far enough out into the corona to determine the value and location of the maximum temperature.

A huge bright coronal condensation appeared over an active region at the limb of the Sun during the 1952 eclipse, and its emission line spectrum was recorded successfully. F. Q. Orrall showed how to extract the three-dimensional temperature distribution of such an object by assuming that it was cylindrical, and analyzing the brightness distribution of the emission lines over its surface. He used the ionization method described above. The condensation turned out to have a hot core, at 3 to 4 million degrees Kelvin, and a cooler halo.

Condensations have been observed and studied in this way at many other eclipses since 1952. Astronomers began to realize that the really hot, dense regions of the corona concentrate above active regions, where the magnetic fields are probably strongest. We will return to this connection between density, temperature, and magnetic field strength a little later in this chapter.

The overall temperature distribution of the corona remained elusive. Astronomers wanted to know the value and location of the maximum temperature in the corona, in order to check hydrodynamic theories of the expansion of the solar wind, but there seemed to be no adequate way to determine this important quantity.

Then in 1970, a new idea dawned. A consortium of English astronomers, including the brilliant theoreticians A. Gabriel and C. Jordan, launched an ultraviolet spectrograph with a rocket and intercepted the umbral shadow of the moon during the eclipse of March 7, 1970. Their spectrograph included in its wavelength range the strongest emission line emitted by hydrogen, Lyman alpha, at a wavelength of 1216 A. The spectrograph was slitless and produced the usual monochromatic images of the chromosphere and corona (see Figure 5.6). The scientists expected to see the corona imaged in spectrum lines characteristic of highly ionized elements, but they were astounded to find an image of the corona in Lyman alpha.

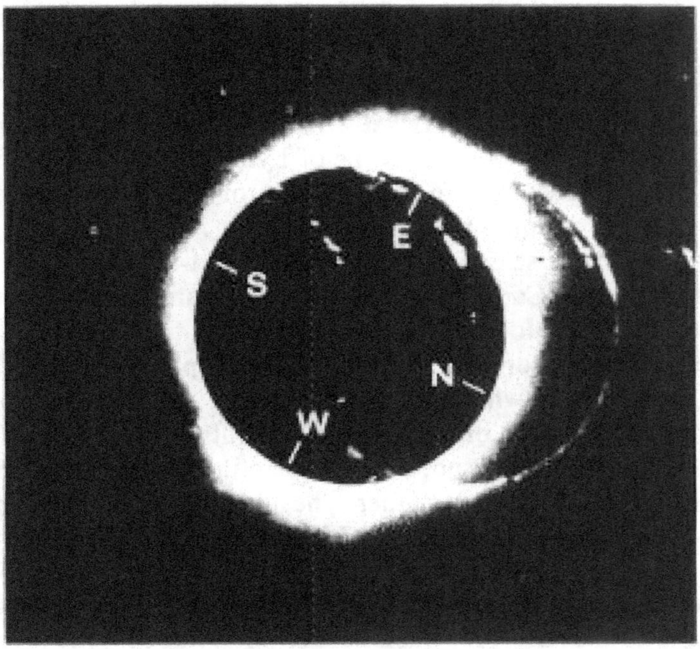

Fig. 5.6 The corona photographed in the UV line of hydrogen (Lyman alpha, 1216 Å), from a rocket during the March 7, 1970 eclipse. The faint image to the right is another nearby monochromatic image in the slitless spectrum. (*Courtesy of A. H. Gabriel, "Measurements on the Lyman Alpha corona," in* Solar Physics *21 (1971), by permission of D. Reidel Publishing Co., Dordrecht, Netherlands.*)

Think a moment what this means. Nearly all hydrogen atoms will ionize, to form protons and electrons, at a temperature of 10,000°K. Yet here was evidence of un-ionized hydrogen atoms in the corona, where the temperature is well over a million degrees. The theorists rushed to their computers, and discovered something they might have realized long before. Only one hydrogen atom in a million manages to survive the blistering heat of the corona, and to retain its precious electron, but because the corona is nearly pure hydrogen, enough neutral atoms of hydrogen survive to scatter the strong Lyman alpha radiation from the chromosphere to the observer in detectable quantities.

Now hydrogen is the lightest of all gasses, so the coronal hydrogen atoms move faster, on the average, than any other element. As a result, the Lyman alpha line of hydrogen is broadened appreciably by the Doppler effect we described earlier. J. M. Beckers and E. Chipman pointed out that the Lyman alpha line provides a superb thermometer for measuring the coronal temperature up to a height of a solar radius or more. Beckers and H. Argo from Los Alamos Scientific Laboratory outfitted a rocket with a Lyman alpha spectrograph, and attempted to observe the profile of Lyman alpha throughout the corona at the eclipses of 1976 and 1980. Unfortunately, both attempts were frustrated by equipment failures.

A consortium of scientists from the Harvard-Smithsonian Center for Astrophysics in Cambridge, Massachusetts, and the High Altitude Observatory in Boulder, Colorado, picked up the challenge. They designed a special coronagraph for Lyman alpha, a Lyman alpha spectrograph, and a white light coronagraph. All three instruments were flown on a rocket on April 13, 1979, outside of eclipse.

Their experiment was successful, and returned data on a coronal hole, a quiet region, and a streamer. The line widths in the quiet region decreased at increasing heights, and indicate a steady decrease in the kinetic temperature with height from 2.6 million degrees K at 1.5 solar radii, to 1.2 million degrees K at 4 solar radii. Since the temperature is decreasing in this range, the maximum temperature must lie below 1.5 solar radii. Unfortunately, their coronagraph was designed to block out all light from heights below 1.5 radii. They concluded, however, that the maximum coronal temperature exceeds 2 million degrees K and lies below 1.5 solar radii.

In the coronal hole, the line width indicates a temperature of 1.8 million degrees K at 2.5 radii. The line width in the hole may be broadened by solar wind expansion as well as by thermal motions, so the actual temperature is apt to lie below the measured value. This type of experiment has a tremendous promise for the future, particularly if the coronagraph can be eliminated and the instrument flown during an eclipse in order to sample coronal temperatures in the vicinity of the temperature maximum.

Much work remains to be done on the temperature structure of streamers. A group of astronomers at the Los Alamos Scientific Laboratory have been photographing streamers from high-altitude jet aircraft during several eclipses. So far, they have observed the 1970, 1972, 1973, and 1980 eclipses. Figure 5.7 shows a computer-enhanced image processed from photographs of the 1980 eclipse. You can trace a number of streamers out to 13 solar radii, and polar plumes out to 8 solar radii. In addition, an unusual eruption was photographed in progress.

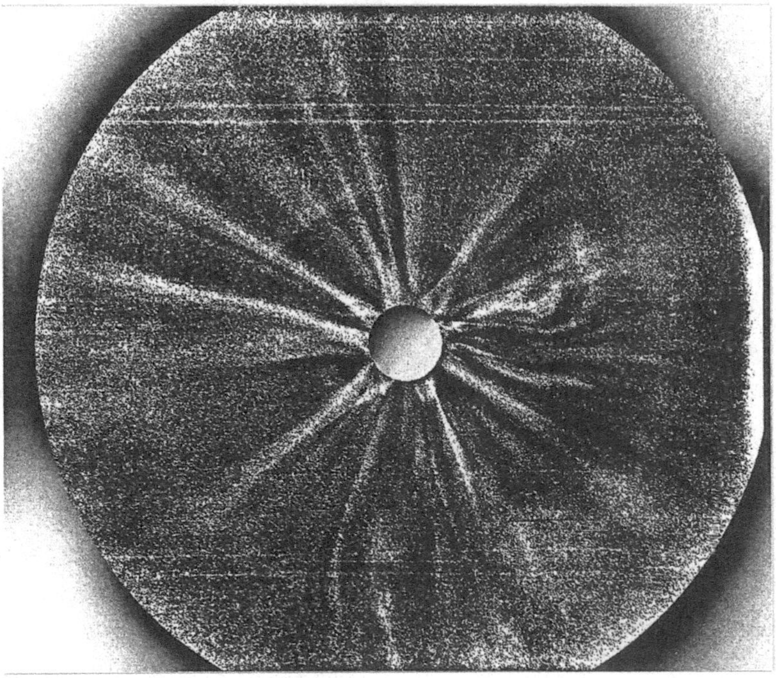

Fig. 5.7 Coronal streamers, photographed from an aircraft during the February 16, 1980 eclipse. The streamers can be traced out to 12 solar radii. The twisted object at 11 o'clock is a coronal eruption which was in progress during the eclipse. (*This photo by C. F. Keller and W. Matuska and reproduced by permission from Los Alamos National Laboratory.*)

Now, although densities can be determined directly from such photographs, temperatures cannot. Since the gas in streamers is not in hydrostatic equilibrium, but flows outward, a simple scale height temperature would be meaningless. Therefore, the Los Alamos group devised a second experiment to measure streamer temperatures.

Remember that the white light of the corona originates in the photosphere, and is scattered toward the observer by coronal electrons. Photospheric light contains Fraunhofer lines, but the fast-moving coronal electrons smear out those lines by the Doppler effect when they scatter the light. Thus, the coronal spectrum is a relatively smooth continuum, showing no evidence of the Fraunhofer lines, with *one* exception. The Fraunhofer lines of ionized calcium, at the wavelengths of 3933 A and 3968 A are the strongest in the visible spectrum. They block out almost 30 A in the photospheric spectrum. They are so strong, that even the Doppler broadening coronal electrons cannot completely smear them out. There remains in the coronal continuum a broad, shallow depression that is the remnant of the combined ionized calcium lines in the incident photospheric light.

When Grotrian first observed this depression in 1939, he immediately grasped its significance as pointing to a high-temperature corona, and even estimated the temperature from the depth of the depression. Since then, astronomers have tried to measure the depression acccurately, and so estimate the coronal temperature.

The Los Alamos scientists repeated this experiment in their jet aircraft in 1980. They obtained no results because of a crucial computer failure. However, the method should be sensitive enough to determine streamer temperatures out to considerable distances from the Sun, and will be repeated in the future.

WHY IS THE CORONA HOT?

The Sun's temperature falls continuously, from 15 million degrees Kelvin at its center to about 6000 degrees Kelvin at its photosphere. The temperature then rises abruptly to reach about 2 million degrees K in the inner corona, and then again slowly declines toward the Earth, where the temperature of the solar wind has been measured at about 100,000 degrees K.

Why is the corona so hot? Why does the temperature jump abruptly beyond the solar photosphere? These questions have teased solar astronomers for 40 years, and despite an enormous amount of work and many beautiful theories, we still don't know definitely why. Some things are quite clear, though. Heat flows from high to low temperatures; that is an unavoidable law of nature. Thus heat cannot flow from the photosphere to the corona. Some other form of energy besides heat or light must be flowing out of the Sun to maintain the corona's high temperature.

In the early 1940s Ludwig Biermann, the distinguished German astrophysicist, proposed that the photosphere emits sound waves which travel upwards, dissipate as shocks, and heat the chromosphere and corona. Photospheric motions are turbulent, and turbulent flow produces sonic noise. (Anyone who has stood near the exhaust of a jet engine can testify to this fact!) Noise, by definition, consists of a collection of sound waves of many different frequencies, unlike the pure tone of a tuning fork. Sound waves propagate at a definite speed, which depends on the gas temperature. As a sound wave moves upward through an atmosphere its energy remains approximately constant. However, because the gas density decreases with height, the amplitude of the internal motion of the wave must increase if the wave energy remains constant. At some height, the speed of internal wave vibrations becomes comparable to the forward speed of the wave itself. At that point, the smooth transfer of energy in the wave from one point to the next is interrupted. The wave turns into a shock, in which all physical quantities, such as temperature, pressure, and density change extremely rapidly from place to place. Dissipation occurs because of viscosity (the friction of one gas layer against another) or by heat conduction or radiation (the loss of the wave energy to its surroundings). The shock dissipates the energy that the wave was carrying.

A long line of distinguished theorists picked up Biermann's idea and developed it mathematically. Among these are M. Schwarzchild, E. Schatzmann, D. Osterbrock, M. Kuperus, and P. Ulmschneider. Everyone agrees that the photosphere emits more than enough sound wave energy to heat the chromosphere and corona. The question is whether any sound waves will reach the corona. Will they form shocks and dissipate lower down, in the chromosphere, instead of in the corona? To test this whole idea, observers want to detect some effect of sound waves in the corona. What should they look for? In particular, what sound frequencies are expected?

Until recently, theorists have been relying on a theory developed by M. Lighthill to describe sonic noise from jet aircraft. This theory predicts that photospheric motions will generate sound waves with periods between 5 and 50 seconds. But not all these waves will reach the corona. In general, waves with longer periods will die off first. In order to predict the height at which a given period or frequency dissipates, one needs a good theory of shock formation that includes all the effects of dissipation. We are still awaiting an adequate theory.

Needless to say, the observational astronomers did not wait patiently for the theorists to come up with an accurate prediction. They tried to detect sound waves in the corona at every frequency and by every technique they could imagine. Coronal sound waves have been sought by their effects in x-rays, ultraviolet light, visible light—including lines and continua—infrared light, and radio waves. The most general statement we can make today is that no waves have been detected with certainty by *any* technique, at *any* wavelength, at *any* sound frequency.

At the 1970 eclipse in Mexico, the High Altitude Observatory made a motion picture of the white light corona, with the hope of discovering rapid changes in coronal forms. No such changes were found then, but they do exist. Two years later, H. U. Schmidt and W. Wagner decided to re-examine the eclipse movie for evidence of coronal waves. They reasoned that a sound wave consists of periodic fluctuations in density which should show up instantly as periodic fluctuations in brightness. They analyzed the film exhaustively, averaging groups of frames and summing larger and larger areas to improve the accuracy of their brightness measurements. They would have been able to detect changes of only a few percent in brightness, but found nothing of this size. They pointed out that their negative results did not necessarily rule out the presence of sound waves in the corona. They would have to be very lucky indeed to observe a nice, clean, periodic fluctuation of brightness, because any line of sight through the corona would, in general, pass through waves that were moving in different directions and were all out of step. These waves would all tend to cancel one another.

The best place to look for a simple wave pattern is in an isolated coronal structure, such as a loop. J. Harvey, of the Kitt Peak National Observatory, tried exactly this at the African eclipse of 1973. He took a rapid series of photographs through a filter that isolates the coronal green line at 5303 A. Harvey's experiment, like all the others in west Africa, was marred by a sandstorm that blew up during totality. Nevertheless, he was able to detect coronal loops at the limb of the Sun in his photographs. Careful analysis of the brightness of these loops showed, once again, no evidence for periodic brightenings.

Because Harvey's experiment was carried out under less than ideal conditions, J. Pasachoff of Williams College decided to repeat the experiment at the 1980 eclipse in India. He also intended to measure brightness variations in green-line loops, and his equipment incorporated two novel improvements. First, he sampled only a small portion of a loop a few arcseconds in diameter, using fiber optics. In this way, he hoped to avoid averaging out the signal from many overlapping waves that cover a large area in the corona. Secondly, he decided to detect brightness fluctuations with a photomultiplier tube, which is intrinsically much more sensitive than photography.

During totality, he cooly selected the most promising green-line loop at the limb, positioned his fiber optics, and recorded his signal. His records do show fluctuations in brightness, but they are not periodic, and are probably caused by inhomogeneities in the Earth's atmosphere ("seeing").

The solar photosphere oscillates with a period of five minutes, and there is more than enough power in these oscillations to heat the overlying atmosphere. Astronomers have wondered since the discovery of these oscillations whether they had anything to do with heating the corona. The chances are not good, because the oscillations correspond to sound waves *trapped* under the photosphere. However, if only a small portion of this trapped energy leaks out and propagates upward, it might heat the corona.

Since most eclipses do not last as long as five minutes, these oscillations would be difficult to detect during the eclipse. T. Tsubaki, from Tokyo University, did the next best thing in 1977 by searching for five-minute oscillations in the corona with a 16-inch coronagraph at Sacramento Peak Observatory. He measured the intensity profile of the green line of Fe XIV (5303 A), as well as its total brightness, and found a curious result. The *width* of the line appeared to oscillate with a five-minute period, but the total *brightness* remained relatively constant. These results are really puzzling, and have not been confirmed.[1]

While the observers are having a hard time finding waves of any kind in the corona, the theorists are merrily constructing heating models of the corona. It has become fashionable to consider not merely sound waves but magnetohydrodynamic waves that propagate along or transverse to the magnetic fields in the corona. As we saw earlier, we know next to nothing about the configuration or strength of such fields, and so the theorist's imagination remains relatively unconstrained for the moment.

Orbiting Solar Observatory Eight, launched in 1975, carried an ultraviolet spectrometer built by the University of Colorado. R. Athay, O. White, and their colleagues observed Doppler shifts of the spectrum lines of ionized silicon at a wavelength near 1838 A and derived the characteristics of chromospheric oscillations. They found that the chromosphere oscillates with periods between 30 seconds and 180 seconds, and determined that this wave energy decreases at increasing heights. They concluded that nearly all the oscillatory energy dissipates at heights below 2000 km in the chromosphere. There is not enough energy left above this height to account for the strong chromospheric radiation, much less for the coronal energy losses.

Their investigation finally persuaded R. Rosner, at the Smithsonian Astrophysical Observatory, that there really *are* no waves heating the corona. In this respect, Rosner is playing the role of the small child in Grimm's fairy tale who declares that the emperor wears no clothes.

Rosner and his colleagues proposed instead that the corona is heated with electrical currents. They pointed out that photospheric motions will twist magnetic lines of force that thread upward into the corona. Such a slow twisting and untwisting generates electrical currents. The solar corona is an excellent conductor of electricity (about as good as sea water), but it does have some finite resistance. Moreover, a completely ionized gas like the corona does not conduct electricity as a copper wire would. Instead, electrical currents generate magnetic fields that tend to pinch the current into narrow channels. The current does not take the "path of least resistance," that is, through the large volumes of the corona, but instead flows through the path of *greatest* resistance! As a result, the current heats narrow filaments in the corona very effectively. In this way, Rosner and his colleagues tried to explain why filamentary and loopy structure prevail throughout the corona.

Their problem now is to find a way to redistribute the heat generated in such filaments throughout the corona. While it is true that the corona exhibits a broad range of temperatures and is extremely inhomogeneous, nevertheless, on the whole it is hot. Heat must somehow move across magnetic lines of force to warm the entire body.

At the present time, Rosner's provocative ideas on anomalous current heating of the corona are simply that—ideas. Experimentalists have yet to work out critical observational tests that would confirm or reject his ideas. No doubt these will come along in the next few years, and some experiments will be made at future eclipses.

MOTIONS IN THE CORONA

We have already described the search for wave motions in the corona. What other kinds of motions exist? There are rotation, expansion, condensation, and eruption. This litany might give you the impression that the corona boils with activity, but in fact a random snapshot-

proves otherwise. Except at special times and special places, the corona is relatively quiescent. Coronal speeds lie in the range of 1 km/sec or less, and are correspondingly difficult to measure.

In order to measure coronal rotation, for example, one needs to follow some long-lived structures, such as streamers or filaments, month after month. Eclipses are entirely unsuitable for this kind of investigation. Most of what we know about coronal rotation we have learned with coronagraphs, such as the High Altitude Observatory's K-coronameter in Hawaii, from radio maps, and preeminently from a nine-month series of coronal photographs obtained with a coronagraph aboard Skylab.

Figure 5.8 displays the coronal rotation rate at different latitudes. As you can see, the Sun's photosphere rotates *differentially,* i.e., more rapidly at the equator than at the poles. At any latitude coronal structures seem to rotate more slowly than the photosphere underlying them. One of the great surprises Skylab provided is that coronal

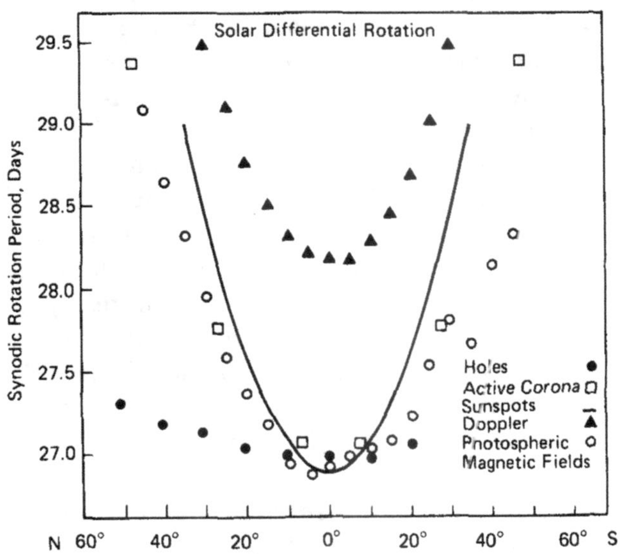

Fig. 5.8 Photospheric and coronal rotation rates, versus solar latitude. Most features rotate more rapidly at the equator than at the poles. Active regions in the corona and photospheric magnetic field structures share the same rotation curve as sunspots (the solid line). However, coronal holes (+) rotate almost as fast near the poles as near the equator. (*from E. Antonucci, "Rigid and Differential Rotation of the Solar Corona," in* Solar Physics, *34 (1974), by permission of D. Reidel Publishing Co., Dordrecht, Netherlands.*)

holes do not rotate differentially, but almost rigidly, as though they were painted on a hard ball. We do not understand the reasons for any of this yet.

We have seen in Figure 5.7 that streamers can be photographed out to 13 solar radii, but as yet no actual measurements of their streaming motion has been possible. Nevertheless, by combining measurements of their density distribution with some theory, astronomers can make some estimates on their rate of outward expansion. The French scientist S. Koutchmy, for example, obtained photographs of the white light intensity and polarization of a large streamer during the March 1970 eclipse.

Using a technique developed by K. Saito in Tokyo to interpret the intensity and polarization measurements, Koutchmy reconstructed the location and shape of the streamer in three dimensions. The streamer rests on two huge "feet" that emerge from large photospheric areas of opposite magnetic polarity. The feet straddle a prominence, seen at the limb. The streamer itself tapers down in thickness with increasing height to a narrow neck some three radii above the limb, and then begins to fan out again (see Figure 5.9).

Koutchmy assumed that the outward expansion velocity within the streamer accelerates rather slowly (i.e. is virtually zero at the base of the streamer) and reaches the speed of sound at the narrow neck. These ideas derive from Parker's theory of the coronal wind, but were also influenced by the resemblance of the streamer to a Laval nozzle. Such a nozzle has a converging section, a narrow neck, and an expansion section. Gas under high pressure in the converging section accelerates rapidly, reaches the speed of sound at the neck, and then expands supersonically in the diverging portion. Koutchmy first determined a scale height temperature of 1.35 million degrees Kelvin from his density measurements. He next assumed this temperature extends all the way up to the neck, where the velocity of sound would be 100 km/sec.

Knowing the density, area, and speed of sound at the neck, he then estimated the mass flux through the streamer, and knowing the density and cross-section in the diverging portion of the streamer, he estimated an expansion speed of 300 km/sec at 6 solar radii. These results are reasonable, but as you can see, depend upon a chain of assumptions. One would really like to know, independently, the temperature and velocity distribution in the streamer.

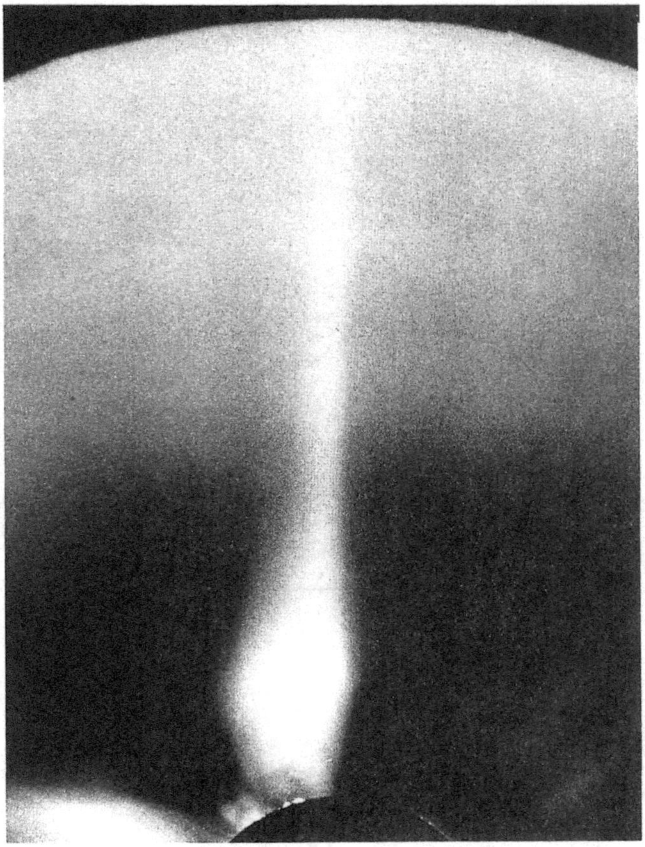

Fig. 5.9 A large isolated coronal streamer, photographed by S. Koutchmy at the March 7, 1970 eclipse. See text for Koutchmy's analysis. (*from Serge Koutchmy, "Etude Hydrodynamique du Grand Jet Coronal Observé à l'Eclipse du 7 Mars 1970,"* in Solar Physics, *24 (1972), by permission of D. Reidel Publishing Co., Dordrecht, Netherlands.*)

G. Pneuman and R. Kopp at the High Altitude Observatory have constructed theoretical models of streamers to try to understand how the characteristic converging, diverging form originates. Figure 5.10 is drawn from one of their papers. The coronal magnetic field is, of course, one of the prime factors involved. A magnetic field exerts a lateral pressure that is proportional to the square of the field strength (B^2). In the lower portion of the streamer, the pressure of the magnetic field exceeds the gas pressure and prevents the gas from expanding laterally. The only escape route for the gas is upward, and it ac-

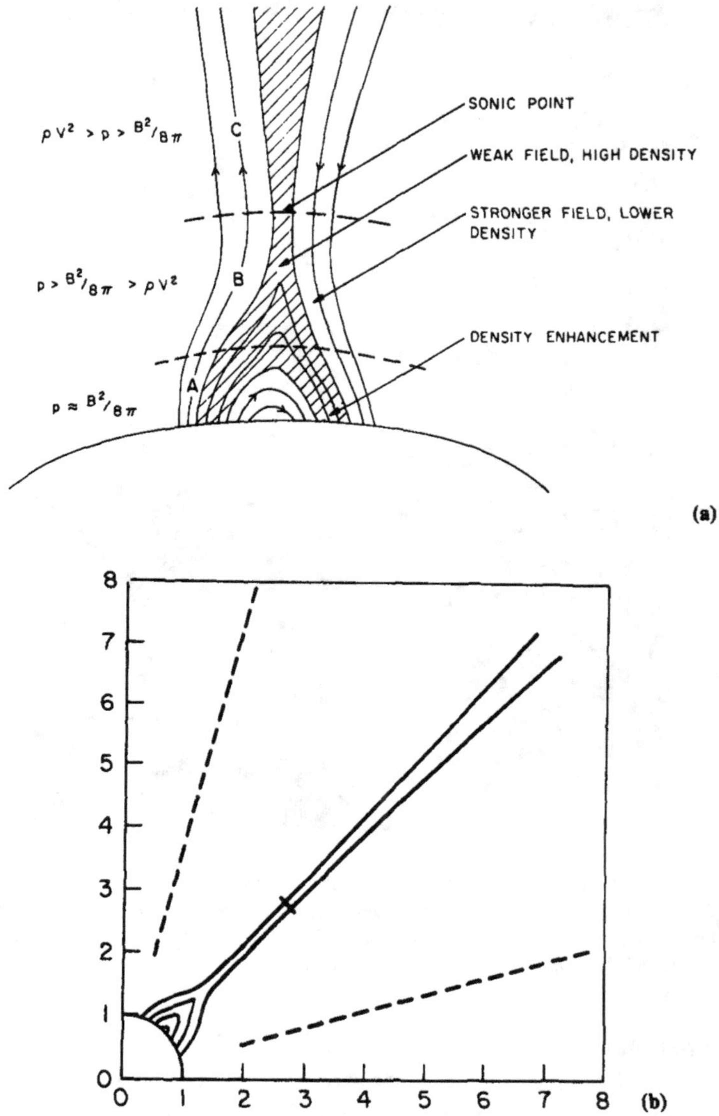

Fig. 5.10 **(a)** A sketch of a typical streamer. The streamer straddles a system of low coronal arches, and extends into the paper as a long, flat fan. The direction of the magnetic field embedded in the streamer reverses polarity at its mid-plane. The gas reaches the speed of sound at the "sonic point."

(b) The calculated magnetic field in a typical open streamer. The sonic point lies at 3 solar radii. *(Courtesy of G. W. Pneuman, "Coronal Streamers," in* Solar Physics, *13 (1970), by permission of D. Reidel Publishing Co., Dordrecht, Netherlands.)*

celerates to reach sound speed at the neck. At this point the magnetic and gas pressures are comparable. Above this height, the pressure and energy of the flowing gas are sufficient to bend the magnetic field lines outward.

Notice that the two halves of the streamer in this model are separated by a thin sheet of electrical current. Kopp and Pneuman introduced this device as a convenience in their calculations, but is it possible that such a sheet exists, and as Rosner has suggested, contributes to the heating of the streamer? Further observations will be needed to test this idea.

The expansion of the wind from coronal holes has been studied intensively. A coronal hole at the north pole of the Sun was observed continuously for two weeks with the white light coronagraph aboard Skylab. R. Munro and B. Jackson analyzed their white light intensity and polarization measurements in much the same way that Koutchmy analyzed his streamer. The observations yield the spatial distribution of electrons, and the cross-sectional area at every height. Munro and Jackson assumed that a definite quantity of gas, (Q), typical of coronal holes, flows through this particular hole. Knowing the density (d) and cross-sectional area (A) they can then figure out what the axial speed (V) of the flow must be from the simple equation $V = Q/dA$.

They discovered that the wind speed increases to 200 km/sec below a height of 2.5 solar radii, and then continues to rise to 450 km/sec at 5 solar radii. Near the Earth, the velocity in such a solar wind stream might be 600 to 800 km/sec. Munro and Jackson concluded that most of the acceleration to this high speed occurs very close to the Sun, and this result furnishes an important clue on how the gas must be heated and accelerated. Nevertheless, future experiments must decide what are the actual temperature and flow speed distributions in a hole.

Motions in prominences have been measured at eclipse, and in great detail with coronagraphs outside of eclipse. The gas streams predominantly downward and relatively slowly. However, prominences are often so large and so dense, that even a downward speed as small as 1 km/sec implies, through the above equation, an enormous flow of gas, and yet the prominence may persist for many months. Where does all this material come from?

Our best guess is that coronal material cools and flows inward into the body of the prominence, but this only shifts the problem. Simple calculations show that a single large prominence could drain the entire corona in ten days or so, unless the corona itself were replenished. Where does the coronal material come from?

We really do not know, although the chromosphere seems the best possibility. In particular, the chromospheric spicules, first seen by Airy and studied exhaustively with coronagraphs through the 1960s and 1970s, could easily supply the necessary amount of coronal material, but we do not definitely know that they do.

If the coronal gas heats up and rises in the spicules and cools down and falls within prominences, we should be able to detect a slow circulation. Several astronomers have looked for this effect at eclipses.

J. Harvey and W. Livingston of the Kitt Peak National Observatory used a special spectrograph to record Doppler motions in the corona at the eclipses of 1970, 1973, and 1980. Their spectrograph was fitted with five parallel entrance slits, which cut across the image of the corona and so sampled coronal motions along five parallel lines. In addition to the corona spectrum, they recorded the spectrum of a neon lamp. The many narrow lines in the neon spectrum serve as a wavelength reference for Doppler shifts of the coronal line.

Livingston and Harvey found, much to their surprise, that the coronal line was *red-shifted* everywhere in the corona. That is, it appeared that the corona was receding from them at all three eclipses! Now this is a very odd result. It is not easy to decide whether a red shift, for example, indicates upward or downward moving material; one has to use auxiliary information on the orientation of coronal structures. In the inner corona where loops prevail, predominant downward motion might be expected, but at the great heights Livingston and Harvey observed, *both* upward and downward motion should be seen. As of this writing, they are still struggling to understand their results. In any case, it is too early to decide whether their findings support the picture of a corona that condenses into prominences.

Several experimenters at the 1980 eclipse tried to measure coronal motions. A group from the United States, headed by R. Smartt and a group from India, led by J. N. Desai, both used Fabry-Perot interferometers to record the intensity and profile of the coronal green line at many different points in the corona. The Indian group hoped to infer variations in the kinetic temperature of the corona from the line width. The American group was more interested in observing Doppler motions in low-lying loops.

Neither experiment has so far detected a systematic circulation, however. They will try again at the Indonesian Eclipse of 1983.

It is time to summarize. Astronomers are gradually converging on the answers to the four major questions posed by the solar corona. They face a difficult assignment, because the corona appears more inhomogeneous, more rapidly changing, more subtle with every advance in experimental or instrumental technique. Still, astronomers are a patient breed, who enjoy a stiff challenge from nature. They will continue to probe the corona's behavior until their curiosity is satisfied.

NOTES

1. S. Koutchmy, Y. Zugzda, and V. Locans have very recently announced (in *Astronomy and Astrophysics*, in press) the discovery of periodic doppler shifts of the Fe XIV line at 5303 A. Periods of 80 and 43 seconds appear in their data. These scientists suggest that the oscillations are Alfven waves in coronal arches.

SUGGESTIONS FOR FURTHER READING

Books

Brandt, J. C. *Introduction to the Solar Wind*. W. Freeman Publishers: 1970.

Parker, E. N. *Interplanetary Dynamical Processes*. Wiley & Sons: 1963.

Zirker, J. B. (ed.) *Coronal Holes and High Speed Wind Streams*. Colorado Associated University Press: 1977.

Periodicals

Kuperus, M., Ionson, J. and Spicer, D. "Theory of Coronal Heating," in *Annual Review Astronomy and Astrophysics* **19**, 7, 1981.

Svalgaard, L. and Wilcox, J. M. "A View of Solar Magnetic Fields, the Solar Corona, the Solar Wind, in Three Dimensions," in *Annual Review Astronomy and Astrophysics* **16**, 429, 1978.

Vaiana, G. S. and Rosner, R. "Recent Advances in Coronal Physics," in *Annual Review Astronomy and Astrophysics* **16**, 393, 1978.

6
Interplanetary Dust

If you face the western sky during twilight on a clear autumn evening, you will see a long triangular streak of light along the Sun's apparent path in the sky (the ecliptic). The Italian astronomer Cassini first noticed this phenomenon in 1683 and correctly attributed the light to sunlight that has been scattered from small solid particles that cluster in the plane of the ecliptic. It is now called the zodiacal light (see Figure 6.1).

Meteor showers are another phenomenon that indicate the presence of solid material in the solar system. The sudden streak of light we call a meteor is emitted by a pinhead-sized body (a "meteoroid") as it enters the Earth's atmosphere at 20 to 80 kilometers per second and burns up. Some meteoroids are big enough to survive this trip and end up in museums. You may have seen one at some time—a huge, pitted, stony or metallic lump, with sharp edges and a distinctly alien look about it.

We now know that the Sun and planets move through a cloud of solid particles that range in size from microscopic dust to chunks of rock and ice three meters across. Astronomers want to know where all this matter came from and what becomes of it. An entire branch of astronomy, called "meteoritics," is devoted to these kinds of questions.

In this chapter we will focus on only the smallest particles—the dust. We will first sketch current ideas on its origin and evolution. Then we will turn to recent eclipse experiments designed to determine the range of sizes, the composition, and the shape of the dust particles. As usual, we will find that space experiments have greatly extended and amplified the information available from eclipses.

Fig. 6.1 The zodiacal light, photographed at twilight from Mount Haleakala, Maui, Hawaii. (*Courtesy of J. L. Weinberg; photo by P. B. Hutchison, January 1967.*)

COMETS AND METEORIC DUST

Astronomers think that the Sun and its planetary system condensed from an interstellar gas cloud. Such clouds typically contain large quantities of molecular hydrogen, some helium, and a sprinkling of all the other metals and nonmetals that appear in the Sun. In addition to cold gasses, interstellar clouds contain solid grains (or "dust") that constitute perhaps 6% of their total mass. The composition and structure of the grains, as you can well imagine, are difficult to determine. They seem to be tiny particles of hard substances, such as magnesium and aluminum silicates, silicon carbide, or graphite. They are probably coated with layers of the ices of such light gasses as water, carbon monoxide, carbon dioxide, cyanogen, and ammonia. The particles can be as small as a few hundredths of a micron (a micron is 40 millionths of an inch).

No one really knows whether the dust that we detect in the solar system has survived unchanged since the Sun formed from its interstellar cloud. Astronomers do think that the most pristine, undisturbed material in the solar system resides in comets.

The Sun has an enormous family of comets, perhaps a billion of them, that are thought to swarm in a great cloud that extends to the nearest stars. Occasionally a comet from this cloud will be pulled into the inner solar system by the gravitational attraction of Jupiter and Saturn. A comet initially consists of a fluffy mass of ice crystals and silicates, and has been compared to a large, dirty snowball.

As a comet approaches the Sun, it heats up. Some of the ice melts, and the volatile gases spew off to the rear, creating a sensational comet tail. Comets may have gaseous tails, dusty tails, or both.

Much of the solid material within our solar system (excluding the asteroids, which are quite another matter) originates in comets. The orbits of meteoroids help to prove this statement. Several times a year the earth experiences showers of meteors, which have been given such names as the Perseids, the Quadrantids, and the Geminids. We know that these meteors are caused by particles that circle the Sun in orbits that are identical with those of their parent comet.

Comets eject dust particles of micron or submicron size as well as the millimeter sized meteoroids. The dust particles also go into orbits around the Sun, but these orbits become slowly unstable. As the dust particles reflect sunlight and radiate heat, they experience a slight drag on their orbital motion. (This process has been called the Poynting-Robertson effect.) As the particles slow down, they spiral inward toward the Sun. Eventually they begin to evaporate, and when they reach some minimum size, perhaps a tenth of a micron, the pressure of solar radiation is sufficient to stop their inward spiraling and to push them out of the solar system. As a result, dust-free zones must exist around the Sun.

This picture of the evolution of dust in the solar system has been accepted as a good working hypothesis, but there are many details that are missing or uncertain. In the remainder of this chapter we will describe some eclipse investigations aimed at testing parts of this overall picture.

THE F CORONA AND THE ZODIACAL LIGHT

As Cassini discovered, dust in the solar system concentrates toward the plane of the ecliptic, that is, the plane in which the Earth moves around the Sun. As a result, sunlight is scattered in a narrow band, centered on the ecliptic, that is called the zodiacal light. The dust particles, as we shall see, extend into the solar corona. The light they scatter there (called the "F corona") is superimposed on the light scattered by coronal electrons.

Astronomers would like to determine the spatial distribution of the dust particles, their size distribution, and the variation of the size distribution with distance from the Sun. By measuring the color and polarization of the zodiacal light, they can learn a great deal about these properties of the dust. If there is any variation in the size distribution, it should be most apparent near the Sun, where the intense solar radiation will tend to evaporate the dust particles and sweep them out of the solar system. Therefore, astronomers want to make measurements of the zodiacal light as close to the Sun as possible.

During a solar eclipse, the sky (as seen from the ground) is at least 1000 times darker than in normal daylight, and the F corona can be traced out to about 5 solar radii. After a normal sunset, the zodiacal light can be detected as close as 20°, or 40 solar radii, to the Sun. The really difficult observational problem lies in filling in the gap between 5 and 40 solar radii. The best approach toward filling this gap has been to fly equipment in high altitude aircraft during a total solar eclipse. From an altitude of 30,000 feet, the sky brightness drops by another factor of 10 and the zodiacal light can be measured readily to great distances.

The British astronomer, D. E. Blackwell, and the French astronomer, R. Michard, were the first to undertake such aircraft experiments, beginning with the eclipse of 1954. These high altitude measurements were supplemented by a whole series of ground-based eclipse observations into the early 1970s. These efforts yielded a definitive map that shows the brightness and polarization of the average zodiacal light versus distance from the Sun.

Now as we have mentioned, coronal electrons and dust in the ecliptic plane along the line of sight both scatter sunlight. The astronomer has somehow to separate these two contributions and to isolate the fraction due to dust alone. There are two ways to go about this.

The first, and most popular, is to *assume* that the dust does not polarize sunlight but that coronal electrons do. By measuring the degree of polarization at various distances from the Sun, astronomers are then able to calculate the fraction of the light that has been scattered by dust and, indirectly, the amount of scattering area attributed to the dust in every cubic centimeter along a radius from the Sun. This method works quite well because the basic assumption, (i.e. negligible polarization by the dust) turns out to be quite reliable. However, until 1966 the assumption had not been tested.

D. E. Blackwell and A. D. Petford used a second, independent method, which we will describe in a moment, to separate the dust and electron components of the inner zodiacal light and to show at what distance from the Sun the polarization due to dust begins to be appreciable.

Blackwell's method is based on the realization that the spectrum of sunlight scattered from coronal electrons contains no Fraunhofer lines. The rapid thermal motion of the hot electrons blurs out the lines and presents the observer with a featureless continuum. On the other hand, the spectrum of sunlight scattered from dust particles *does* show the Fraunhofer lines because, apparently, the random motion of the dust particles in their orbits is small. Figure 6.2 illustrates the effect of adding these two spectra.

If there were no electrons in the corona, and only dust scattered sunlight, we would see Fraunhofer lines with the same shape and depth as in ordinary sunlight, but since the electrons contribute an essentially flat spectrum, the observed lines are shallower. The ratio of the central intensities in the scattered light and in the incident sunlight is a measure of the amount of dust along the line of sight.

Blackwell and Petford built an instrument that incorporates these principles. They chose to measure the depth of the hydrogen line, at 6563 A in the red part of the solar spectrum. Light from the center of the line and from the nearby continuum was presented, alternately, to a single photomultiplier tube. The difference between the electrical signals gave the desired depth of the spectral line. They flew their instrument in an aircraft at 30,000 feet over Great Slave Lake, in Canada, during the total solar eclipse of July 20, 1963.

Their measurements of the total brightness and polarization of the

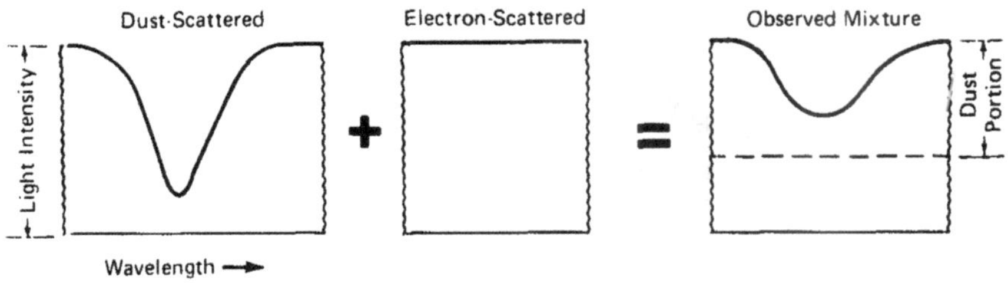

Fig. 6.2 When dust scatters sunlight, the scattered spectrum contains Fraunhofer lines, but when electrons scatter sunlight, their motion obliterates the lines. When *both* scatter sunlight, the resulting spectrum is a mixture. The lines appear, but are shallower in depth than in photospheric light.

zodiacal light and F corona extended through the interesting region bewteen 5 and 16 solar radii from the Sun. The polarization introduced by dust turned out to be quite small indeed in this region. It varied from .1% at 8 radii, to .8% at 16 radii. For comparison, the polarization reaches 20% at 150 radii. These results validate the usual method of separating electron from dust scattering in the corona that uses polarization measurements alone.

Blackwell and Petford were able to separate very accurately and definitively the contributions of electron and dust scattering to the total intensity of the F corona and zodiacal light. The two contributions are equal at about 2 radii from the center of the Sun. The electron contribution falls off more rapidly and reaches only 4% of the dust contribution at 16 solar radii (see Figure 6.3).

Blackwell and his associates next faced the task of deriving the spatial distribution and size distribution of the dust from measurements of the brightness, color, and polarization of the F corona and the zodiacal light. The method they used had been developed some 30 years before by H. van der Hulst and C. W. Allen. The trick is to represent the number, n, of dust particles at any distance, r, as a power law, namely $n = n_o(R/r)^k$ and the number N of dust particles with diameter d as $N = N_o(d_o/d)^l$. Moreover, one has to assume how an individual dust particle scatters light in different directions—the so-called "scattering function."

Since the only known rigorous theory of scattering applies to spheres or ellipsoids, one must assume, tentatively, that the dust has these shapes. Finally, one has to guess the composition and the structure of dust particles. Astronomers have tried various assumptions,

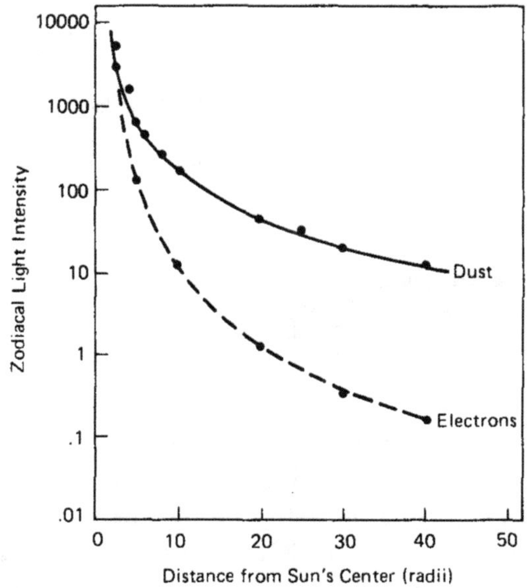

Fig. 6.3 The intensity of sunlight scattered by dust and by electrons, versus distance from the Sun. The separation of these two components was accomplished by Blackwell and Petford.

such as dielectric materials (like ices), metals (like iron), or some combination, such as ice-coated grains. If all these assumptions are folded into a mathematical theory, one can calculate the brightness, color, and polarization of the zodiacal light as a function of distance from the Sun, and compare these with observations.

By adjusting the constants of the theory, such as the indices *k* and *l* in the above equations, one can improve the fit. Blackwell and his colleagues found that the best fit to the brightness and color data between 2 and 50 radii implied a value of *k* in the spatial distribution that lies between 1.0 and 1.5, and the value of *l* in the size distribution between 4 and 5. In other words, the *amount* of dust decreases away from the Sun, and the smallest sizes (about a micron) are most numerous. However, *no* choice of the constants enabled Blackwell and his colleagues to reproduce the observed variation of linear polarization. Their calculations always fell short of the observed value, which suggests that their assumptions about the shapes and composition of the dust particles were wrong.

Blackwell's aircraft measurements represented the high point of zodiacal light observations during eclipse. Further improvements in measuring the brightness, color, and polarization of zodiacal light have been made with a series of spacecraft experiments, including Skylab, Heios 2, and Helios 1 and 2. Satellite experiments have the advantage of observing zodiacal light against a perfectly black sky, without any contamination from the Earth's atmosphere.

R. H. Giese and E. Grun analyzed the Heios 2 observations in 1976. They found that they could reproduce the observations of brightness and color by assuming a power law size distribution with an index *l* equal to 2.2. Particles in the range between 5 and 100 microns were responsible for 80% of the zodiacal light. As we shall see later, their results are open to question, even though they may account for the observations, because Giese and Grun assumed *spherical* dust particles and a density of 3 grams per cubic centimeter, which is now known to be much too large.

Observations of the brightness, color and polarization of the zodiacal light from Helios 1 and 2 have given the most definitive estimate so far of the spatial distribution of dust particles. These two satellites ranged inward from the Earth's orbit (1 astronomical unit, or AU) to .3 AU. C. Leinert and his colleagues at the Max Planck Institute of Astronomy in Heidelberg determined a power law index *k* of 1.3 ± .1 from the change in the brightness of the zodiacal light as the satellites approached the Sun. The color and polarization of the zodiacal light remained constant over this distance range, which indicates that the distribution of sizes does not change over this portion of the solar system.

To date, Leinert and his colleagues have made no attempt to estimate the size distribution from these zodiacal light measurements. The reason, probably, is that the best estimates of sizes now come from other space experiments.

These are of three types. Some satellites like Heios 2, Pioneer 8 and Pioneer 9 carried sensitive detectors to *count* the flux of interplanetary dust particles that strike the satellite. Many techniques are used, including sensitive microphones, semi-conductors, ionization cells, capacitors, photocells, and thin wires. The second technique involves counting *microcraters* on lunar rocks. Craters as small as a fortieth of a micron have been measured on some of these rocks. With suitable calibration, the distribution of crater sizes can be converted into the size distribution of incident particles. The third method involves actually *collecting* dust particles, from high altitude balloons or spacecraft.

Figure 6.4 shows distribution of masses of dust particles in the Earth's vicinity as derived from lunar crater statistics and the Heios 2 dust experiments. If we assume that the density of particles is 1 gram per cubic centimeter—about that of water ice—we can convert the masses into equivalent diameters, and these are shown on the bottom scale of the figure. The curve can be represented by three different power laws, with an index $l = 2.2$ in the interesting range of 5 to 100 microns. Giese and Grun used these data in their 1976 interpretation of the zodiacal light.

In all the investigations we've cited so far, astronomers have tacitly assumed that interplanetary dust consists of spherical, or at best, needle-shaped particles, and have differed mainly on their composition and physical structure. It came as quite a shock, therefore, to actually *see* an interplanetary dust particle. D. E. Brownlee and his col-

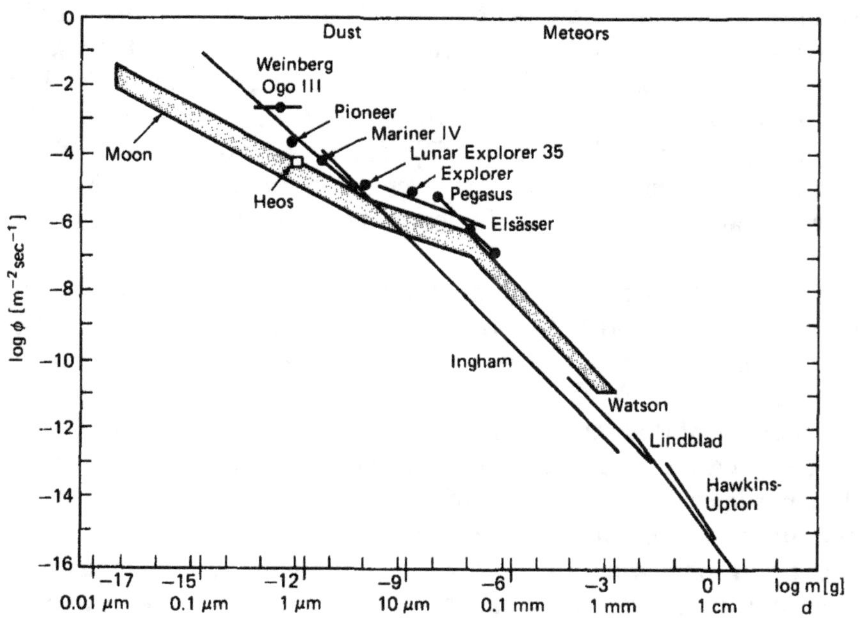

Fig. 6.4 The mass (or size) distribution of interplanetary dust particles. Most particles are smaller than 0.1 micron. Particles around 1 micron in diameter produce the zodiacal light. (*Courtesy H. Fechtig, Max-Planck-Institute für Kernphysik, Heidelberg.*)

leagues at the University of Washington collected "micrometeorites" in the stratosphere from a U-2 aircraft. Most of the particles they collected were essentially undamaged.

Figure 6.5 is a picture of one, made with a scanning electron microscope. The bar at the bottom of the picture is 10 microns long. As you can see, the particle has a "fluffy" structure, very different from a sphere. It consists of submicron grains cemented in a porous matrix. Later on in the chapter we will return to the question of the chemical composition of such particles.

When astronomers first saw such fluffy particles in 1976, they realized that no existing theory would predict the intensity and polarization of light that they scatter. They would have to carry out

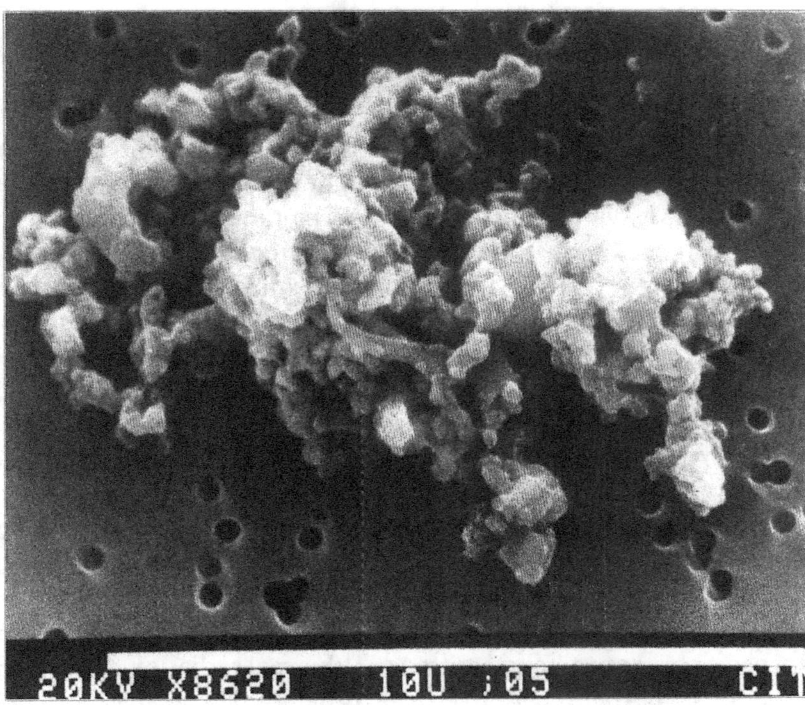

Fig. 6.5 A "fluffy" micrometeorite collected in the stratosphere. The photograph was made with a scanning electron microscope. The length of the white bar represents 10 microns. *(from D. E. Brownlee, et al., "Analysis of Interplanetary Dust Collections," in* Solid Particles in the Solar System, *proc. of IAU Symposium 90, ed. by Halliday and McIntosh; copyright © 1980 by the International Astronomical Union.)*

laboratory experiments instead. Following the suggestion of J. Mayo Greenberg, they built centimeter sized models, illuminated them with millimeter radio waves, and measured the intensity and polarization of the scattered radiation.

The scattering properties of fluffy particles depend primarily on their composition. Dielectric particles hardly polarize incident radiation at all, and even this small amount is virtually independent of the scattering angle. In contrast, absorbing particles (e.g. metals) polarize incident light strongly with a peak in the vicinity of 70 to 80 degrees. Both dielectrics and absorbing particles scatter more radiation in the forward and rear directions than they do to the sides.

Scientists at Ruhr-University Bochum have shown that all the observed properties of the zodiacal light can be explained by the scattering properties of inhomogeneous fluffy particles that consist partly of dielectrics and partly of absorbing grains.

CHEMICAL COMPOSITION OF THE DUST

As early as 1929, Henry Norris Russell, the eminent American astronomer, recognized that zodiacal light particles would evaporate near the Sun, leaving a dust-free shell. The radius of the shell is the nearest distance a dust particle can approach the Sun without evaporating entirely, and thus is an indication of its evaporation temperature, or composition. No terrestrial material can survive closer than two or three radii from the Sun. How could astronomers hope to detect the presence or absence of the dust this close to the Sun?

In 1963 Allen W. Peterson suggested a way. He calculated the thermal emission and the scattered radiation of dust. We need to distinguish between these. The surface of the Sun has a temperature of about 6000°K and the light it emits has a spectrum that is characteristic of this temperature. Dust particles orbiting the Sun receive this light, absorb some of it, and reflect or scatter the rest. The absorbed energy must be re-emitted as "thermal radiation," i.e. heat, or the dust particle will overheat rapidly. Thus, the particle's equilibrium temperature is determined by its absorbing, scattering, and emitting properties as well as its distance from the Sun. Sample calculations showed Peterson that these temperatures are in the neighborhood of a few hundred or a thousand degrees. Most of the thermal emission lies in the infra-red part of the spectrum. Peterson found that the thermal emission exceeded the scattered radiation at wavelengths longer than about 2 or 3 microns.

Near the edge of a dust-free shell, the dust is as hot and its thermal radiation is as intense as it ever will be. Peterson showed that the edge of a dust-free shell should show up as a slight bump on the distribution of infra-red light in the F corona. He put his predictions to the test at the total eclipse of November 12, 1966.

Peterson set up his equipment, a 6-inch f/4 telescope with a lead sulfide detector, near the tiny, bleak village of Huachacalla, on the alto plano of Bolivia. This location, at an altitude of 13,000 feet in a dry climate, was ideal for infra-red observations. Peterson recorded the intensity of the F corona during the eclipse at a wavelength of 2.2 microns on both sides of the Sun. His records showed a small bump at 4 solar radii in the tracing.

He was also able to observe the corona at 3.5 microns, after the eclipse. He saw once again a peak on each side of the Sun at 4 solar radii. The observed emission at 2 microns was less than he had calculated, while that at 3.5 microns was about right. He concluded that dust extends to at least 4 radii from the center of the Sun. The material of which it is composed must evaporate between 1000° and 2000°K.

R. McQueen, a student at Johns Hopkins University, carried out a very similar experiment at the same eclipse at Pulacayo, Bolivia. Although the duration of totality there was only 93 seconds, McQueen recorded a nice bump in the emission at 2.2 microns at a distance of 3.75 solar radii from the Sun's center. Figure 6.6 shows his tracing.

A year later, McQueen repeated the experiment from a balloon, flying at 28 km above Palestine, Texas. During a 5 hour run he was able to confirm the emission bump, which now lay at 4.1 radii, and found two new bumps at 8.7 and 9.2 radii.

The high evaporation temperature implied by these two results suggested a refractory material like quartz as a possible composition for the dust particles that reach this close to the Sun. However, without more spectral information, it was not possible to narrow down the composition.

Peterson therefore repeated his experiment at the total eclipse of March 7, 1970 near Nejapa, in the state of Oaxaca, Mexico. This time, he recorded the intensity of the emission at 4 radii at three

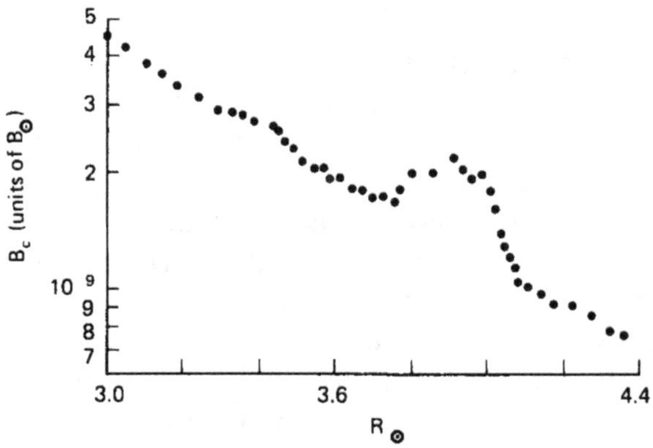

Fig. 6.6 MacQueen's observation of infra-red emission from interplanetary dust at the 1966 eclipse. (*Reprinted courtesy of Robert H. MacQueen and* The Astrophysical Journal, *published by the University of Chicago Press; © 1968 The American Astronomical Society.*)

wavelengths: .84, 1.57, and 2.23 microns. The relative intensity of the emission at these three wavelengths determined a vaporization temperature of 2160° ± 200°K. Such a very high vaporization temperature pointed again to a highly refractive material, possibly graphite.

Peterson tried once again at the eclipse of June 30, 1973, at Chinguetti, in Mauritania. Unfortunately, a sandstorm ended just before the eclipse and the air was so heavily laden with dust that it transmitted only 12% of the infra-red light from the Sun. Peterson's experiment was therefore frustrated.

An international team led by Pierre Lena, of the Paris Observatory, had much better luck during the same eclipse. Lena and his colleagues, Y. Viala, A. Sufflot, and D. Hall of the Kitt Peak National Observatory in Tucson, carried their experiment aboard the Concorde 001 supersonic aircraft. This superb machine raced down the track of totality, keeping pace with the Moon's shadow. As a result, these four experimenters experienced a 74 minute totality.

They made scans along the ecliptic between 3 and 19 solar radii at the east limb of the Sun. Their detector recorded the emission at a wavelength of 10 microns. Figure 6.7 shows their results. They saw bumps not only at 4 radii but also at 12, 13, and 17 radii.

They also measured the spectrum of the F corona for the very first time between 8 and 13 microns. At a distance of 3 radii, their spectrum

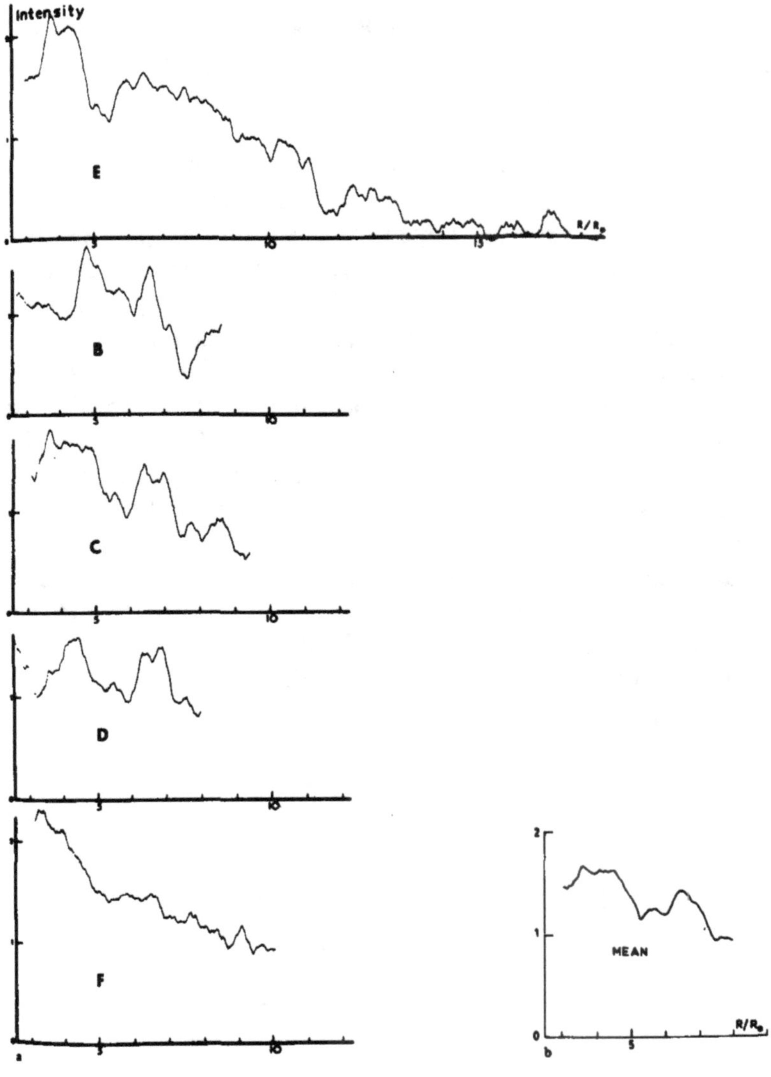

Fig. 6.7 The infra-red emission of dust particles within 15 radii of the Sun. Observations obtained by P. Lena from the Concorde, during the 1973 eclipse. (*from P. Lena et al., "The Thermal Emission of the Dust Corona, during the Eclipse of June 30, 1973. Part I. Instrument Design and Performance,"* in Astronomy and Astrophysics, *37 (1974), and through the courtesy of the European Southern Observatory.*)

showed a strong absorption feature at 10.8 microns and an emission feature at 9.5 microns (see Figure 6.8). These two features are known from laboratory experiments to show up in the spectrum of silicates. Lena and his colleagues speculated that the dust this close to the Sun may consist of glass particles.

Their spectrum also showed an increasing intensity beyond 11 microns. This rise in the infra-red emission ruled out such materials as graphite, but it still leaves left open the possibility of quartz particles.

Meanwhile, astronomers at the University of Minnesota became interested in looking for iron carbide grains. Edward Ney, a prominent infra-red astronomer from that university, had been investigating the appearance of dust shells in a variety of astrophysical objects. Such shells show up several months after the eruption of a nova. For example, Nova Serpentis, Nova Vulpeculae, and Nova Cygni all developed dust shells after their ejecta cooled sufficiently. Dust shells also appear in some kinds of carbon-rich stars, like R Corona Borealis and in Wolf-Rayet stars. Ney found no evidence of an absorption or emission feature at 10 microns that might indicate the presence of silicates. Instead, the spectrum of these dust shells was smooth, and its overall shape indicated a dust temperature of about 1000°K.

Collaborating with John Lewis, a physicist at MIT, Ney investigated the temperatures and pressures at which a number of can-

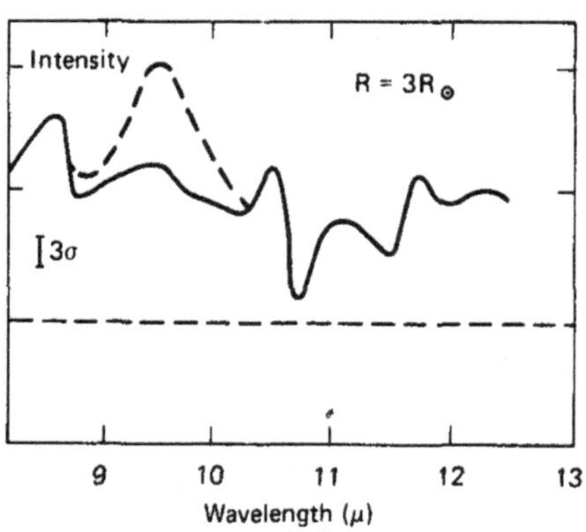

Fig. 6.8 The infra-red spectrum of dust, at a distance of 3 solar radii from the Sun, during the 1973 eclipse. (*from P. Lena et al., "The Thermal Emission of the Dust Corona, during the Eclipse of June 30, 1973. Part I. Instrument Design and Performance,"* in Astronomy and Astrophysics, *37 (1974), and through the courtesy of the European Southern Observatory.*)

didate materials for dust would condense. Iron carbide (Fe_3C) looked like the most likely possibility.

Inspired by these results, Ney resolved to search for iron carbide in the F corona of the Sun. He led an expedition to Bowbells, North Dakota, to observe the February 26, 1979 eclipse. His experiment consisted of a telescope which looked, alternately, in a 0.25° beam above and below the Sun at an angular distance of 3°. His observations were made at a wavelength of 3.5 microns.

Ney found a brightness peak at 24 solar radii, which is just where you would expect iron carbide to evaporate. Unfortunately, there were some cirrus clouds in the sky during the eclipse, and they might have produced the observed infra-red bump.

Therefore, Ney repeated his experiment at the total eclipse of February 16, 1980 at Japal-Rangapur, in India. Once again, clouds were present, and contaminated the 3.5 micron signal from the corona. Thus, the presence of iron carbide remains uncertain. Ney might try again at the Indonesian eclipse of 1983.

By 1979, most astronomers were convinced that the infra-red bump at 4 solar radii in the corona was emitted by some form of silicate dust. However, K. Brecher, of Boston University, and his associates, posed a very different explanation for this feature. They pointed out that not only Saturn, but also Uranus and Jupiter have rings around them. They wondered whether the Sun might possess its own ring of solid particles, and suggested a variety of chemical and physical reasons why it might.

Carbon, or iron carbide, they estimated, could survive as close as 4 solar radii in the form of massive chunks, 10 km in radius. These large bodies would scatter sunlight, and should appear as 8th to 10th magnitude "stars," but would be difficult to see close to the Sun, even during eclipse, and might easily have been missed.

A group of astronomers from the Indian Satellite Center in Bangalore, led by U. R. Rao, decided to search for this hypothetical ring around the Sun during the Indian eclipse of 1980. They scanned the equator of the Sun between 2 and 5.5 solar radii, recording the radiation at 2.2 microns. To everyone's surprise, they did not detect a bump at 4 solar radii. They claim this null result rules out the ring of graphite objects proposed by Brecher. However, they also had to conclude the F-corona varies in brightness, since several observers *had* seen a bump at 4 radii before 1980, and they had not. This conclusion raises questions of its own, since satellite observations of the zodiacal light indicate that it is very stable, and does not vary.

The most direct and convincing way of investigating the chemical composition of interplanetary dust, is actually to analyze some in the laboratory. As we noted above, D. E. Brownlee and his colleagues from the University of Washington, collected interplanetary dust in the stratosphere from a U-2 aircraft. In 1979, they reported on the composition of these particles.

A few of the particles (less than 40%) are single, solid mineral grains, usually of iron sulfide, or silicates like forsterite or enstatite. The single particles are usually encrusted with a fine-grained opaque material which has solar elemental composition. Single grains of calcite and magnesium phosphate have also been collected.

The great majority of the collected particles are black, and have the "fluffy" structure shown in Figure 6.5. They consist of a large number of grains that range in size from 0.01 of a micron to several microns. These grains are cemented together with an amorphous material that appears to be different from any known meteoric material. The submicron grains embedded in the matrix are usually not composed of a single mineral. They are mixtures of minerals such as olivine and pyroxene, along with iron and nickle sulfides. Some of the fluffy particles are very porous, and have a mass density of much less than 1 gram per cubic centimeter. Others are more compact, with densities of 2 or 3 grams per cubic centimeter.

Presumably, these micron sized-particles were originally coated with ices of light gasses, but these ices have not survived entrance into the stratosphere. So, although we have yet to see a dust particle as it exists in its parent comet, we can confirm that the dust contains the silicates (but not the quartz or graphite) that are deduced from astronomical observations. Much work remains to reconcile the laboratory analyses with the infra-red dust spectra.

THE ORBITAL MOTION OF THE DUST

Theoreticians have worked out the evolution of the orbits of interplanetary dust under the influence of solar radiation, collisions, and the Poynting-Robertson effect. The picture they give us is, in broad outline, the one described above. The Poynting-Robertson drag causes a particle's orbit to shrink, so that it gradually spirals in toward the Sun. As the particle approaches the Sun, it absorbs radiation strongly, and the pressure of this radiation tends to counteract the Poynting-Robertson effect. As a result, the perihelion distance (the closest approach to the Sun) stabilizes, but the orbit becomes increasingly elliptical until, finally, radiation pressure pushes the particle out of the solar system. The size of the particle is then about a tenth of a micron. Although its outward velocity is less than 20 or 30 km/sec relative to the Sun, the Earth encounters it at speeds of something like 40 km/sec.

Experimentalists have attempted to check the predictions of the theorists in two ways. They either attempt to measure the speed, direction, and mass of particles in space using equipment on spacecraft, or they try to measure particle speeds using spectroscopic methods from ground-based experiments during a total eclipse.

To date, four spacecraft have given us information on motion of interplanetary dust. These are Pioneer 8 (in heliocentric orbit between 1968 and 1972), Pioneer 9 (in heliocentric orbit between 1969 and 1972), Heios 2 (geocentric orbit from 1972 to 1974), and Helios 1 (heliocentric orbit from .3 to 1.0 A.U. from 1974 to 1979).

Pioneer 8 and 9 and Helios 1 detected solid particles streaming *from* the Sun. Not much is known about these grains, except that they are small (less than a tenth of a micron in diameter or 10^{-15} grams) and fast (moving at 50 km/sec or more). They fit the picture of particles being swept out of the solar system by radiation pressure. They have been called "beta meteoroids," since beta, in the jargon of the specialists, is the ratio of the forces of radiation pressure and gravity.

The four spacecraft also detected particles arriving from the forward direction of the spacecraft. Since these space vehicles were all revolving in orbits from west to east, the direction of arrival of the dust suggests that it moves in elliptical orbits around the Sun. Experiments aboard Heios 2 determined that the incoming speed of the dust particle lies between 10 and 23 km/sec. These small numbers imply that the particles were caught close to their aphelion (the point furthest from the Sun). These orbiting particles are considerably heavier than the beta meteoroids, and range in mass up to 10^{-11} grams.

Still more massive particles were encountered by Heios 2, in equal numbers from all directions in the ecliptic. These particles are, presumably, the zodiacal light particles that are slowly spiraling inward toward the Sun. The principal conclusion one might draw from spacecraft data so far is that two and possibly three populations of dust can be distinguished by their orbits.

There is still a great deal to be learned, and the prospects for dust experiments on future spacecraft are rather dim at the moment. However, the eclipse method we are about to describe has great potential.

Willett Beavers and his colleagues at the Iowa State University seem to be the only group in the U. S. who are trying to measure the radial velocities of interplanetary dust particles using spectroscopic techniques during an eclipse. They have applied a method which is well known in stellar astrophysics. The idea is to measure the Doppler shift of as many lines as possible in the spectrum of the object you are looking at. To do this without a great deal of complication, you prepare in advance a mask for the focal plane of your spectrograph. The mask has curved slots at many wavelengths where strong absorption lines will appear in the object's spectrum. All the light that passes through the slots is collected and presented to a single photomultiplier.

For the F corona experiment, Willett Beavers prepared a mask by first photographing the spectrum of daylight, which contains many strong Fraunhofer lines. Then, using a special device (a "mask maker") on a machine that is ordinarily used to measure the position and depth of Fraunhofer lines, he prepared a mask with 17 slots at the position of the corresponding lines. This mask was then mounted in the focal plane of his spectrograph; it could be driven back and forth in the direction of the wavelength with a motor.

Now the light of the F corona, as we have seen, contains exactly the same Fraunhofer lines as daylight. However, because dust particles move in the line of sight, the entire spectrum of the light they scatter, including the Fraunhofer lines, will shift, either to the blue or the red. As a result, there will be a slight mismatch between the position of the slot and the position of the incident Fraunhofer line, resulting in more or less light passing through the slots to the photomultiplier. The amount of light received, therefore, can be calibrated to give a measurement of the speed of the dust particles in the line of sight. In Willett's experiment, the slot widths corresponded to 120 km/sec, and a shift of the lines relative to the slot of 20 km/sec would produce a 20% intensity rise.

Beavers and his colleagues first tried out the experiment at the eclipse of February 26, 1979. They took their 10 centimeter, f/7.6 telescope with a little spectrograph to Riverton, Manitoba, on the west bank of Lake Winnepeg in Canada. The totality phase lasted only 2 minutes and 52 seconds.

During this time, the spectrograph slit was placed, west of the Sun, in a north-south direction, at a distance of 3.2 radii from the Sun's center, and then at 4.3 radii. Figure 6.9 shows Beaver's results.

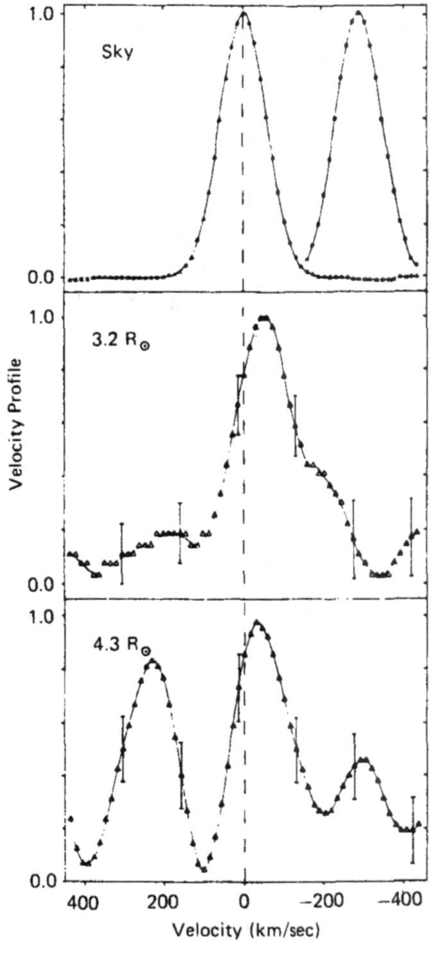

Fig. 6.9 The radial velocity distribution of dust, observed by W. Beavers during the 1979 eclipse. (*Reprinted courtesy of Willet I. Beavers and* The Astrophysical Journal, *published by the University of Chicago Press; © 1980 The American Astronomical Society.*)

At both distances from the Sun, Beavers recorded dust approaching the Earth, i.e. blue-shifted, with a mean speed of 50 km/sec. The velocity profile at 4.3 radii also shows the presence of dust receding from the Earth at 250 km/sec. Curiously, the velocity profile corresponding to 3.2 solar radii shows no sign of such a feature. This result implies to Beavers that inhomogeneities in the interplanetary dust may exist over distances as small as 1 solar radius.

If you examine Figure 6.10, you will see that dust that orbits the Sun in the same direction as the Earth, i.e. "prograde," will produce *red* shifts if it is observed west of the Sun, whether it is on the near side or the far side of the Sun. The red-shifted feature in the record that Beavers obtained at 4.3 radii might be associated with dust in prograde orbits, particularly since the orbital velocities it implies are high enough to be reasonable.

The velocities in the *blue*-shifted features, however, are too small to represent orbital motion at this distance from the Sun. Dust that produces this signature in the record may be moving radially with respect to the Sun, according to Beavers. If the radial motion were the same all around a circle centered on the Sun, then, as Figure 6.10 shows, the velocity components of the dust on the near and far side of the Sun

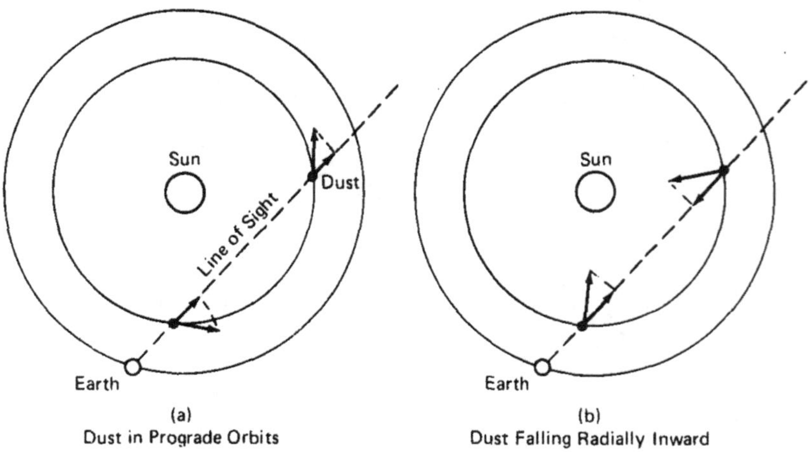

Fig. 6.10 **(a)** Emission from dust revolving in prograde orbits will be red-shifted.

(b) Emission from infalling dust will be red-shifted near the Earth but blue-shifted beyond the Sun. The shifts may cancel.

will tend to cancel, whether the dust falls into the Sun or streams away from it.

The blue-shifted component, therefore, implies either *more* dust or *faster* dust on one side of the Sun than the other. These observations in themselves, however, are insufficient to distinguish between infall and outflow. Once again, they do imply interesting inhomogeneities in the motion of the dust near the Sun.

Beavers repeated his experiment at the Indian eclipse of 1980 in order to clarify his 1979 observations. Unfortunately, a microcomputer that he had installed between the eclipses failed, and Beavers got no data at all. He will try again at the Indonesian eclipse of 1983.

SUMMARY

What do we know at this point? Interplanetary dust is "fluffy" and complex in chemical and physical structure. Small grains (less than a micron) predominate, although grains as large as tens of microns have been collected. The medium range of sizes (around one micron) scatter sunlight most efficiently, and produce the zodiacal light. Despite the infra-red evidence for grain evaporation, the size distribution of grains doesn't seem to vary in the solar system, at least as close as 0.3 AU (or 60 solar radii), to the Sun. Better observations will be needed to detect the influence of the small, evaporating, grains on the F-corona. The overall picture of the particle orbits, and their changes with time and distance, seems consistent with the present space observations, but these can be extended and checked with additional eclipse experiments. We will just have to wait for further developments.

SUGGESTIONS FOR FURTHER READING

Books

Halliday, I. and McIntosh, B. A. (eds.) *Solid Particles in the Solar System*. Proceedings of I.A.U. Symposium 90. Reidel Publishing Company: 1980.

Hodge, P. W. *Interplanetary Dust*. Gordon and Breach: 1981.

Martin, P. G. *Cosmic Dust*. Clarendon Press: 1978.

Periodicals

Brownlee, D. "Interplanetary Dust," in *Reviews of Geophysics and Space Science* **17**, 1735, 1979.

Ney, E. "Star Dust," in *Science* **195**, 541, 1977.

Savage, B. D. and Mathis, J. S. "Observed Properties of Interstellar Dust," in *Annual Review Astronomy and Astrophysics* **17**, 73, 1979.

Whipple, F. L. "The Spin of Comets," in *Scientific American* **242**, 124, 1980.

7
The Earth's Atmosphere

Atmospheric physicists have been studying the effect of the Sun's radiation on the upper atmosphere for many years. Although they have a good general understanding of what goes on there, they are totally in the dark on some of the most important details. They study the upper atmosphere in every possible way, using balloons, rockets, aircraft, and satellites, as well as more conventional ground-based instruments. Although experiments during total solar eclipses are not the most important means of studying the atmosphere, they certainly have contributed their fair share to our present knowledge, and you can readily see why. A total eclipse of the Sun is about as close to a controlled experiment as an atmospheric scientist can hope for. Sunlight diminishes at a uniform and predictable rate, and near totality, the dark umbra of the Moon sweeps across the top of the atmosphere in a narrow and predictable path. Scientists can then study how chemical and physical processes in the atmosphere respond to the absence of sunlight. An eclipse has the advantage over a sunset of occurring very rapidly, and with the Sun hardly moving in the sky.

In this chapter, we will sample some of the research carried out on the Earth's atmosphere during total eclipses of the Sun. Some of the most important advances in this field were made during the 1970 and 1973 eclipses, and we will describe these at some length. The more recent eclipses, of 1980 and 1981 received considerable attention, but it is a little too soon to expect much more than brief reports on the results. We will describe these insofar as they are available.

SHADOW BANDS

Just before second contact and just after third contact, when the exposed crescent of the Sun is very narrow, eclipse observers see light and dark bands of light race across the ground. These are the famous "shadow bands." They were first described after the total eclipses of 1706 and 1820, and have intrigued amateur and professional astronomers ever since. Most recently, meteorological physicists have measured the bands in order to study temperature gradients in the lower atmosphere. However as we shall see, they have not succeeded so far.

R. W. Wood first gave the commonly accepted explanation for the bands in his *Physical Optics,* in 1911. The bands are related to the twinkling of the stars or "scintillation." Light from a star is refracted (i.e. focused and defocused) by the corrugated boundary between warm and cool air masses (see Figure 7.1). The corrugations act as weak positive or negative lenses, which redistribute the starlight into bright and dark patches at the Earth's surface. The width of the patches is typically 3 or 4 centimeters so that one eye may be in a light patch while the other is in shadow. As the wind aloft moves the bumps in the boundary, the patches shift rapidly and the star appears to twinkle.

Shadow bands arise in much the same way. The narrow solar crescent is imaged on the ground by the weak atmospheric lenses, as parallel light and dark bands. The bands are faint, have low contrast, and move rapidly. They also occur, very briefly, at the most exciting phases of an eclipse. As a result, visual observers tend to disagree on their impressions. Even well-organized parties of visual observers, whose main task is to study the bands, will disagree!

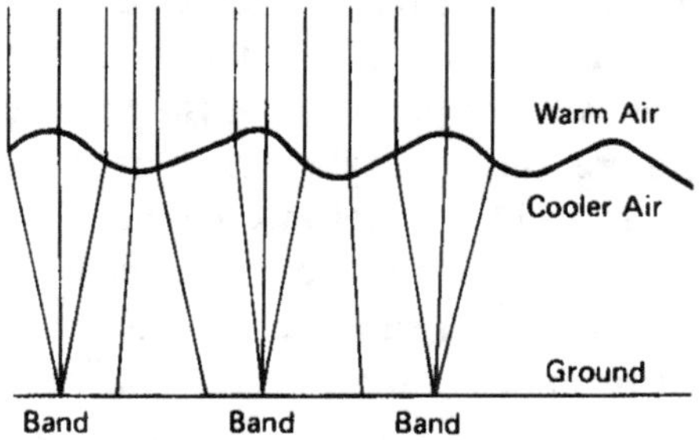

Fig. 7.1. The irregularities in the boundary between layers of warm and cool air act as weak lenses that focus the solar crescent on the ground to form "shadow bands." The effect is similar to the streaks one sees on the bottom of a swimming pool. As the winds aloft move the boundary, the bands move across the ground.

For example, five groups of observers from Ball State University observed the bands during the March 7, 1970 eclipse. The observers were strung out across the path of totality in North Carolina and Virginia. They set up white screens on which clock faces were drawn. Each group recorded the orientation and direction of movement of the bands relative to the clock face. They also timed the passage of the bands across the width of the screen to find their speed. All groups agreed that the bands lie tangent to the umbral shadow or, equivalently, parallel to the chord of the solar crescent, as seen at each location. They also agreed the bands moved *toward* the shadow before totality and *away* after totality. But they disagreed on the spacing (8 to 20 centimeters), the width (1 to 3 centimeters) and the speed (2 to 3 meters/second).

Clearly, visual observations are not accurate enough. Many observers have tried photography, especially with movie cameras, but the results are generally disappointing. A party from the NASA Langley Research Center obtained a very low contrast movie at the 1970 eclipse that showed that the bands are irregular in direction and in length, and move approximately at right angles to their length.

Photoelectric recording of the bands has yielded the most precise information on most of their characteristics. A. Healy, a student at Wesleyan University, recorded the intensity of the bands in blue and in yellow light at the November 12, 1966 eclipse. He found the band width and separation shrank as totality approached and widened after totality. At second contact, the bands were separated by 6 centimeters in blue light, but 8 centimeters in yellow light! The phenomenon was obviously more complicated than anyone realized.

One of the most elaborate photoelectric experiments was carried out by a team from NASA's Goddard Space Flight Center. They arranged photocells at Wallop's Island, Virginia, to record bands in four colors and two polarizations at the March 7, 1970 eclipse. Two distinct types of bands appeared. Their widths were 30 centimeters (or 4 centimeters), and were separated by 200 centimeters (or 8 centimeters) and moved at 16 meters/second (or 3 meters/second). The bands appeared earlier and lasted longer in blue light than in green; blue bands lasted 7 minutes but lasted in green for only 4 minutes. No bands appeared in the infrared, and the bands were unpolarized.

This team confirmed that the bands lie parallel to the chord across the solar crescent and that the bands grow narrower and closer as totality approaches. They noted that the bands travelled from west to east at all times. Radiosonde measurements were made at the direction and speed of the wind, up to a height of 40 kilometers. The wind was from the west at all altitudes.

The NASA group used simple principles of geometrical optics to calculate the height of origin of the two types of bands. Type 1 (with 200 centimeter separation) formed below 3500 meters altitude, while Type 2 (with 8 centimeters separation) formed below 150 meters. The measured wind speeds at these altitudes agreed with the measured speeds of the bands. Thus, the team concluded the bands are caused by "air pockets of different density from that of their environment, made visible by the light of the crescent Sun."

Why do the bands last longer in blue light than in green? The NASA scientists suggested a connection to the shape of the Sun's edge, which is much sharper in blue light than in green, and sharper in green than in the infra-red. This is an interesting idea, but has not been worked out in mathematical detail.

Although the 1970 eclipse yielded definitive photoelectric observations of the bands, two independent Indian groups tried to improve on them during the eclipse of February 16, 1980. A team from Udiapur Solar Observatory observed the intensity fluctuations in the bands photoelectrically at one location. They found that the bands repeated at periods of 3 and 1.7 seconds. These periods are much longer than anyone else has found, and certainly do not correspond with the impression one gets simply by watching the ground. The second group, from the Indian Meteorological Department in Calcutta, measured the width (4.2 centimeters) and separation (14.2 centimeters) of the shadow bands. These numbers agree reasonably well with the Type 2 bands observed by the NASA team in 1970. Note, however, that the width, separation, speed and contrast of the bands may change from one eclipse to another, depending on the size, curvature, temperature difference, and wind speed of the irregularities in the atmosphere.

A comprehensive theory for the bands has not been developed so far. In particular, the narrowing of the bands, their separation, and their increasing contrast as totality approaches, haven't been explained. Moreover the temperature gradients in the atmosphere that are required to produce the bands have not been inferred from the band observations as yet.

SKY BRIGHTNESS, COLOR, AND POLARIZATION

The blue color of the daylight sky arises from sunlight that is scattered by molecules in the Earth's atmosphere. Because the molecules are smaller than the wavelength of visible light, they scatter blue light more efficiently than red, according to a law discovered by Lord Rayleigh. Skylight is not only bluer than sunlight, but is linearly polarized as a result of the scattering process.

The main features of the scattering of sunlight have been understood for many years. S. Chandrasekhar, the eminent astrophysicist at the University of Chicago, devised an elegant mathematical description of the Rayleigh scattering of sunlight in a planetary atmosphere in 1950. This theory predicts the intensity, color, and polarization of skylight for any elevation of the Sun. Elaborate numerical calculations were carried out later by K. Coulson, J. Dave and Z. Sekera, using Chandrasekhar's solution. However, this "classical" description is incomplete in at least two respects. First, it assumes that light is scattered only *once* in the atmosphere and ignores multiple scattering. Secondly, it considers only Rayleigh scattering by air molecules, and ignores scattering by aerosols in the atmosphere. The particles of an aerosol are generally larger than the wavelength of light and scatter according to a quite different law ("Mie" scattering) than Rayleigh's.

The distribution of aerosols in the atmosphere is, of course, variable and it is difficult to account for them in a comprehensive mathematical model. Multiple Rayleigh scattering complicates the model but can be incorporated, once and for all time—at least in principle.

During a total solar eclipse, a large proportion of skylight is multiply-scattered. You can appreciate this if you think of the light that reaches your eye from the dark face of the Moon (see Figure 7.2). If we ignore the light of the corona and air-glow, all this light originates on the hidden disk of the Sun. To reach your eye, it must scatter at least once from molecules in the penumbral shadow of the Moon, and at least once more within the umbra. If you look away from the Moon, the skylight you see is partly singly-scattered from the penumbral shadow, and partly multiply-scattered from within the umbra. The closer toward the horizon you look, the larger is the proportion of singly-scattered light. The proportion of singly- or

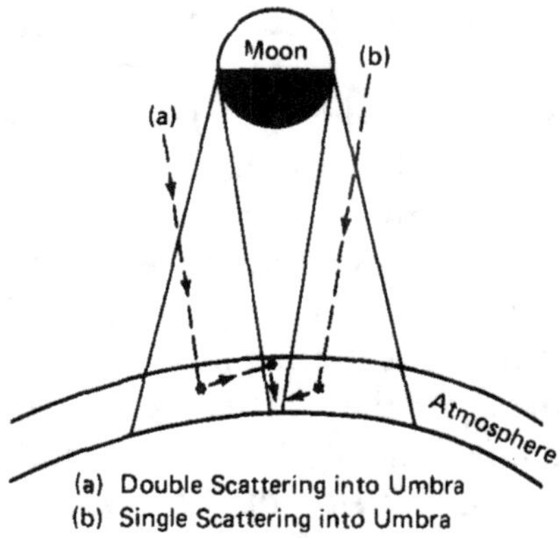

Fig. 7.2. Multiple scattering in the Earth's atmosphere during a total eclipse. The crosses in the figure show where scattering occurs. An observer standing in the umbral shadow sees sunlight that has been scattered twice—as in (*a*), or once—as in (*b*). Higher-order scattering is also possible.

multiply-scattered light also influences the linear polarization of skylight, since multiple scattering tends to *depolarize* skylight. Thus, a solar eclipse offers an atmospheric scientist a fine chance to observe the effects of multiple scattering, and to check his models of the transfer of sunlight through the Earth's atmosphere.

Many experimenters have attempted to measure and interpret the brightness, color, and polarization of the skylight during a total eclipse. We will describe only one successful experiment, that of G. Shaw (University of Alaska) at the June 30, 1973 eclipse. This experiment and its analysis were unusually complete.

Shaw was located at the oasis of Loiyengalani, on the shore of Lake Rudolph, in Kenya. He employed three photoelectric photometers. The first measured sky brightness at six wavelengths, the second measured sky brightness in red light at the zenith, and the third measured the percentage of linearly polarized blue light and red light. During the eclipse, Shaw's instruments scanned the sky repeatedly in two directions: the vertical great circle that passed through the Sun (the "solar vertical plane") and the vertical great circle perpendicular to the first circle (the "perpendicular vertical").

Just before totality, the sky brightness was symmetrical around the Sun, as you might expect, with a bright ring at the horizon (see Figure 7.3). At mid-totality, the intensity of skylight dropped by a factor of nearly 10,000 in the solar vertical, 90° from the Sun. (If you've been to

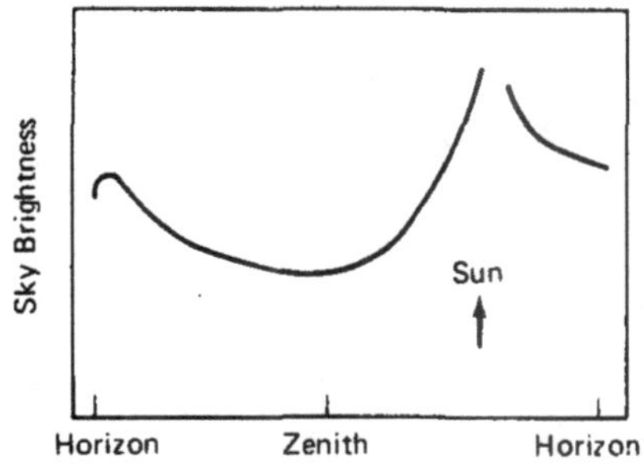

Fig. 7.3. The distribution of sky brightness in the vertical plane that passes through the Sun, just before an eclipse. "Zenith" is the point directly overhead. (*Courtesy of Glenn E. Shaw and* Applied Optics; © *1975, Optical Society of America.*)

an eclipse, you know that the sky at totality is about as bright as during an evening with a full moon). Figure 7.4 shows the brightness of the sky along the "solar vertical plane" and the perpendicular plane during mid-totality. Now the sky brightness is symmetrical about the *zenith,* not the Sun, presumably because the number of scattering molecules decreases toward the zenith.

The sky changes in color, as well as brightness, as every observer knows. Shaw found that the zenith grows much bluer at totality because multiple scattering favors blue light. The horizon, on the other hand, became reddish, because most of the light there was single-scattered into the umbra from the penumbra, and because of the usual absorption of blue light by dust and haze.

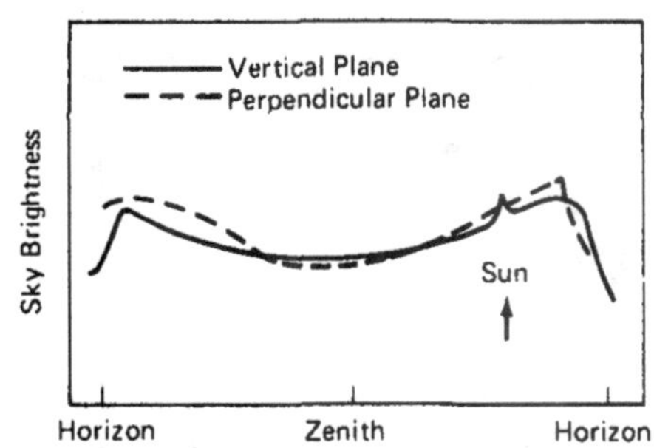

Fig. 7.4. The distributions of sky brightness during a total eclipse in the vertical plane that passes through the Sun (solid line) and in the vertical plane perpendicular to the other plane (dashed line). (*Courtesy of Glenn E. Shaw and* Applied Optics; © *1975, Optical Society of America.*)

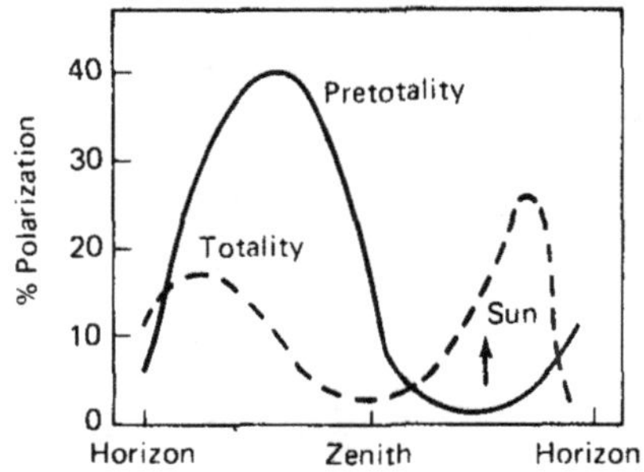

Fig. 7.5. The distribution of linearly polarized light (in percent), before totality (solid line) and during totality (dashed line). (*Courtesy of Glenn E. Shaw and* Applied Optics; © *1975, Optical Society of America.*)

The amount of polarization of skylight is an important clue to the prevalence of multiple scattering, since many scatterings tend to randomize or depolarize the light. Figure 7.5 shows Shaw's experimental results in the solar vertical plane. As with the sky brightness, the polarization becomes symmetrical about the zenith at mid-totality. Notice that, as expected, the percentage of polarized light drops by more than a factor of two. Chandrasekhar's theory predicts that the maximum polarization will appear at 90° from the Sun before eclipse and Shaw's data confirm this.

In 1978, Shaw compared his eclipse observations with an "approximate" mathematical model. The model includes single and double (but not triple and higher-order) Rayleigh scattering, and isotropic scattering by dust. He didn't attempt to calculate the sky's polarization (a ferocious job!) but only the brightness and color. In Figure 7.6,

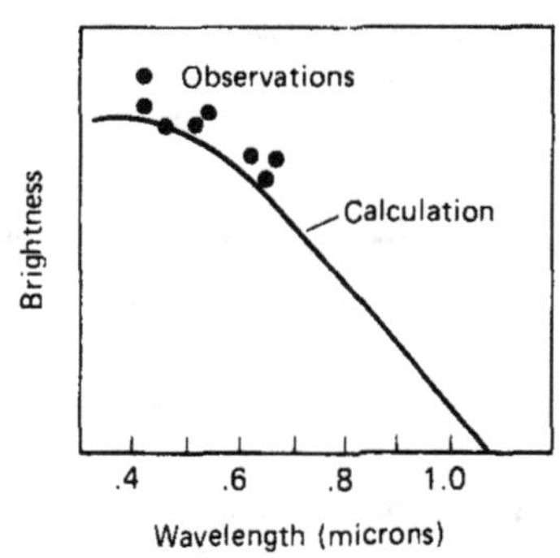

Fig. 7.6. A comparison of Shaw's observations and calculations of the spectrum of skylight at the zenith during totality. (*Courtesy of Glenn E. Shaw and* Applied Optics; © *1975, Optical Society of America.*)

you see Shaw's calculation for the spectrum of the zenith, compared with his observations. The agreement is reasonably good. The model also reproduces the observed distribution of light over the sky, and the variation of color with zenith angle. In particular, the model predicts a blue zenith and a red horizon. Shaw concluded from all this that triple and higher-order scattering accounts for less than 20% of the intensity at any place on the sky, and that the main observable features can be accounted for with single and double scattering.

METEOROLOGICAL EXPERIMENTS

Anyone who has witnessed a solar eclipse will remember how rapidly the air chilled during totality. Some experienced eclipse observers also agree that clouds have a tendency to dissipate as the eclipse approaches totality. In one well-documented case, the 1965 eclipse on Bellinghausen Island, a small hole appeared at precisely the right time and place to allow the ground-based eclipse observers to see the eclipse!

Meteorologists are, of course, interested in the changes in the temperature, density, and wind structure of the atmosphere throughout the total eclipse. The layer of air a few meters thick, that lies directly in contact with the Earth's surface, holds special interest, if only because it is the easiest to measure.

During the 16 February 1980 eclipse in India, Indian meteorologists carried out no fewer than 10 independent experiments designed to measure temperature, density, wind speed, and wind direction, as well as a number of more esoteric quantities, throughout the eclipse. Without exception, they detected the usual cooling trend, but the actual size of the temperature drop (between 1° and 2°C) varied from place to place. At least one observer (R. Narasimham of the Indian Institute of Science in Bangalore) discovered that the air near the ground recovers only very slowly from the eclipse effect; even 4 hours after the end of the eclipse (i.e., after 4th contact), the temperature and average wind speed were still different from their values on a normal day. S. Sethuraman, of the Brookhaven National Laboratory, observed the bottom three kilometers of the atmosphere with many radiosondes. The air throughout the entire thickness of this layer cooled throughout the eclipse.

Although the air at ground level cools, there seems to be some question whether the electrically neutral air at stratospheric heights responds to a total eclipse at all. During the eclipse of 12 November 1966 in Argentina, H. N. Ballard and his colleagues fired a series of 12 sounding rockets in order to measure changes in the temperature, the wind, and the concentration of ozone in the stratosphere that might accompany an eclipse. Their studies showed as expected, that the temperature in the 50–60 kilometer region decreased as the eclipse progressed toward totality, and then recovered after the eclipse to its normal afternoon value.

J. Horvath and J. Theon obtained contrasting results during the March 7, 1970 eclipse. They launched three rockets from Wallop's Island during the eclipse, and, as a control on the experiment, a rocket the day before and the day after the eclipse. The rockets were instrumented to measure pressure, temperature, and density at heights between 30 and 125 kilometers. Their experiments showed that the eclipse had essentially *no* effect on the density profile below 95 kilometers. The three rockets traced out the same density curve to within 1%.

Above 95 kilometers, however, there was a question: the density seemed to decrease by 20% relative to the day before and after. Being careful experimenters, Horvath and Theon returned exactly a year later and fired three more rockets at exactly the same times as during the eclipse. Their results showed that the day-to-day variation of density and temperature are larger than any induced by the eclipse. We will remind you of these results when we describe, later on, the search for waves set up by the cooling of the atmosphere during the eclipse.

VERTICAL STRUCTURE OF THE ATMOSPHERE

We will pause here to introduce some useful background on the structure and composition of the upper atmosphere in order to help you better appreciate the experiments we consider next.

In Figure 7.7, you see a sketch of the height distribution of temperature in the Earth's atmosphere. We live in the troposphere, where the temperature declines with height. About 25 kilometers above us lies the stratosphere. There, the temperature rises very abruptly, starting at about 100 kilometers, and levels off near 1500°K in the thermosphere, which begins around 300 kilometers. The ionosphere is a layer of partially ionized gas beginning at a height of 60 kilometers. It is subdivided into three layers: the D region (60–90

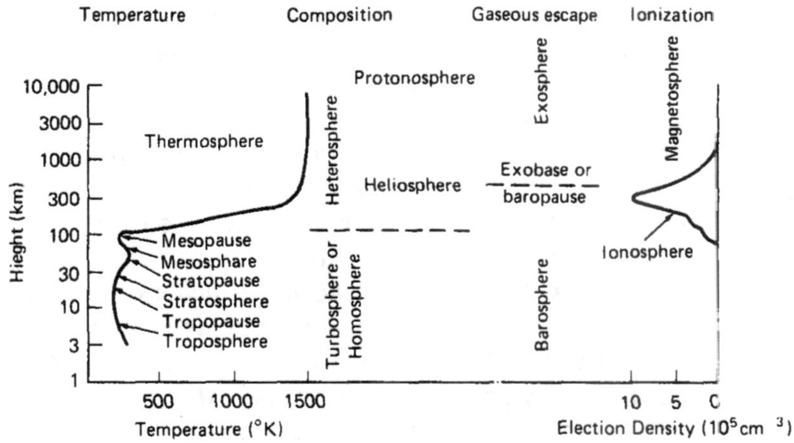

Fig. 7.7. Names of different portions of the Earth's atmosphere, as distinguished by temperature, composition, escape of gas molecules and ionization. (*from* The Upper Atmosphere and Solar-Terrestrial Relations *by J. K. Hargreaves; copyright © 1979 by Van Nostrand Reinhold Co.*)

kilometers), the E region (105–160 kilometers), and the F region (above 180 kilometers).

The curious wiggles in the temperature curve of Figure 7.7 result from the balance of heat gains and losses. The gains are all due to the absorption of solar radiation; between 20 and 50 kilometers, ozone (a molecule composed of three oxygen atoms) absorbs ultra-violet light with wavelengths between 2000–3000 A. If it were not for the ozone layer, life on Earth would be exposed to this harsh, sterilizing radiation. Molecular oxygen absorbs radiation down to about 1000 A at heights below 95 kilometers. Between 70 and 150 kilometers, molecular oxygen, atomic oxygen, and molecular nitrogen absorb ultraviolet light down to about 900 A and soft x-rays with wavelengths between 10 - 100 A.

After a molecule has absorbed a solar photon, storing the energy momentarily, it has three choices: (1) it can re-emit the energy in a different wavelength band (say, the visible or infra-red region), producing a faint light known as the air-glow; (2) if it first suffers a collision with another ion, this ion might come away with more kinetic energy than it started with—in effect, stored energy has been recovered to heat the gas; (3) the stored energy may release one of the molecule's bound electrons, leaving the molecule positively charged. The liberated electron collides very soon with other electrons and ions, because of the long-range electrostatic forces between them, and rapidly distributes its excess energy to the gas. Once again, the original solar photon has served to heat the gas.

Free electrons in the upper atmosphere are constantly being released to form atomic or molecular ions, recombining with such ions to form neutrals, or attaching themselves to neutrals to form negative ions. The ionosphere, for example, contains a rich mixture of positive ions and free electrons. Nitric oxide (whose chemical symbol is NO) ionizes in the D layer by absorbing the strong solar emission line at 1216 A to form the ion NO^+. Similarly, molecular oxygen (O_2) becomes ionized (O_2^+) by absorbing soft x-rays in the D layer.

Higher in the E layer, soft x-rays in the range 10 to 100 A form such ions as N_2^+, O_2^+, and O^+. Still higher in the F layer, where longer-wavelength ultra-violet light is absorbed, such molecules as O_2^+, N_2^+, O^+, He^+, and N^+ appear. The most abundant ions are NO^+ and O_2^+ in both the E and F layers.

The ionosphere does not completely disappear at night, despite its dependence on solar radiation. The concentration of free electrons does, however, decrease to only 10% of its daylight value. There are two reasons why the ionosphere survives the night. First, there is some transport of ions and electrons from the day-lit side of the atmosphere, and second, some ions take a very long time to recapture an electron once they have lost it. As we shall see further on in this chapter, the rates of recombination of ions and electrons for different photochemical reactions are some of the important quantities that atmospheric scientists can determine from total eclipses.

PHOTOCHEMISTRY OF THE UPPER ATMOSPHERE

A total eclipse of the Sun offers atmospheric physicists an excellent opportunity to observe changes in concentrations of different atmospheric ions in response to the cut-off of solar radiation. They are interested in learning which chemical reactions are going on continuously, how rapidly these reactions occur, and in particular, how rapidly a particular ionic species disappears after the solar radiation is cut off. These questions all bear on the processes that heat and cool the atmosphere and that lead to the equilibrium temperature profile shown in Figure 7.7.

L. Brace and his colleagues at the NASA Goddard Space Flight

Center obtained very useful information on these processes during the eclipse of March 7, 1970. They fired two rockets, 30 minutes and 5 minutes before totality, to measure the concentrations of such ions as N^+, N_2^+, NO^+, O_2^+, O^+, and the total ion concentration. At the same time, the rocket instruments measured the temperature of molecular nitrogen and of free electrons. The rockets were fired from Wallop's Island and rose to a height of 290 kilometers into the thermosphere.

The experiment showed that the neutral atmosphere hardly responds to the eclipse, while the ion composition, and particularly the concentration of free electrons, responds quite rapidly. For example, the experimenters found that the temperature and the concentration of neutral molecular nitrogen hardly changes. This result can be understood when you realize that a vertical column of air, 1 cm^2 in cross-section, contains 5×10^4 ergs above a height of 120 kilometers. From the measured temperature profile, the NASA scientists estimated that such a column loses heat by downward conduction only at the rate of .27 ergs per second so that the column would take 51 hours to lose all its energy! In short, such a column of air responds so sluggishly to changes in its energy supply, that its temperature cannot change appreciably during the few hours of the eclipse.

On the other hand, the total concentration of *ions* dropped by 50% during the eclipse in the E layer, and by more than 30% in the lower part of the F layer. The electron temperature dropped from 1700 to 1400°C, almost 20%, in the lower F layer. The rate of cooling of electrons followed very closely the rate of decrease of exposed solar disk throughout the eclipse. This behavior clearly indicated that ions are created and destroyed rapidly, that electrons distribute their excess energy rapidly, and that the x-ray sources that excite the thermosphere are distributed uniformly over the solar disk.

Ozone (O_3) and nitric oxide (NO) are rare, but important, constituents in the upper atmosphere. Ozone absorbs sunlight in the 2000–3000 A part of the spectrum and heats the atmosphere between 20 and 50 kilometers. To form ozone, a molecule of two oxygen atoms must split into atomic oxygen by absorbing a photon of ultraviolet light with a wavelength between 1027 A and 1750 A. An atom of oxygen may then combine with a molecule of oxygen to form ozone. Once formed, ozone absorbs the 2000–3000 A radiation and breaks up again into atomic and molecular oxygen. The cycle repeats continuously.

Nitric oxide heats the D layer by absorbing the powerful hydrogen

Lyman-alpha line in the solar spectrum at 1216 A. NO forms when atomic nitrogen combines with molecular oxygen and can be destroyed to form nitrous oxide (NO_2) when it reacts with ozone. There are several other ways in which both ozone and nitric oxide can be produced and destroyed, however. Eclipse experiments have helped to establish the relative importance of different possible reactions.

Making certain initial assumptions on the dominant reactions, D. Wuebbles and J. Chang, of Lawrence Livermore Labs, and independently, J. R. Herman from NASA's Goddard Space Flight Center predicted the effect that a solar eclipse would have on the stratospheric concentrations of ozone and nitric oxide. Because the concentration of nitric oxide varies strongly during the day, you would expect it to decrease during a solar eclipse. On the other hand, the concentration of ozone should show only minor increases or no changes at all.

To test these predictions and the photochemistry on which they rely, W. L. Starr and his colleagues from the NASA Ames Research Center in California prepared an experiment for the 26 February 1979 eclipse across the northwestern U. S. and Canada. They conducted their experiment aboard a U-2 aircraft, which flew back and forth within the path of totality near Great Falls, Montana, at an altitude of 20 km. For almost three hours before, during, and after the eclipse, they measured the concentrations of nitric oxide and ozone and the air temperature outside the plane. Their experiment showed that the concentration of nitric oxide dropped by at least a factor of 25 during totality! The decrease and recovery of nitric oxide concentrations followed the predictions of the two independent theories quite well. As expected, no change was observed either in the concentration of ozone or in the air temperature.

This experiment also confirmed that nitric oxide disappears in the atmosphere by combining with ozone, and is produced, at least in part, by the break-up of nitrous oxide (NO_2).

Molecular oxygen is one of the major constituents of the atmosphere. The radiation it emits is an important factor in cooling the air in the mesosphere, between 50 and 80 kilometers. Here, molecular oxygen radiates two strong bands in the infrared, at 1.27 microns and 1.58 microns. Before a molecule of oxygen can radiate, however, it must receive and store some energy. Atmospheric scientists are debating exactly how this is accomplished.

The favored theory claims that the destruction of ozone provides that energy. When a molecule of ozone absorbs ultraviolet light around 2700 A, it breaks up into one molecule and one atom of oxygen. The molecule could come away with some internal stored energy, which it later might release as infrared light. This light would contribute to the so-called "day-glow" that is present during all the daylight hours.

In order to test this theory, P. C. Wraight and M. Gadsden, of Aberdeen University in Scotland, designed an experiment for the 30 June 1973 eclipse which was flown aboard the Concord. This supersonic plane flew along the eclipse track nearly as fast as the eclipse shadow, and at an altitude of 17 kilometers. As a result, the experimenters experienced a total eclipse of the Sun for 74 minutes, and a very dark sky against which to measure the faint infrared light. Their instrument was installed in the rear of the plane to look through a starboard window at various angles from the horizon, up to 50° from the zenith.

They discovered that the intensity of the 1.5 micron band *increases* toward the horizon, while the intensity of the 1.2 micron band *decreases*. The two Scottish scientists concluded from these results that ozone is indeed the primary source of excited oxygen molecules that radiate the day-glow.

Their conclusion was not accepted outright, however, Two Indian scientists from the University of Poona (V. Agashe and F. Rathi) decided to repeat the experiment from the ground during the total eclipse of 16 February 1980. They set up their infrared photometer at Gadag, along the eclipse path. Since they did not have the advantage of a 74-minute totality, they had to give up the luxury of scanning the spectrum during the eclipse and of scanning the sky. Instead, they isolated the infrared emission bands with narrow-band optical interference filters and pointed in only a few directions. A cooled lead sulfide detector was used as the sensitive element in their photometer.

Both bands decreased in intensity, throughout the eclipse. If the theory of formation of molecular oxygen by the destruction of ozone were correct, their eclipse observations would imply a *decrease* in the rate of destruction of ozone. At the same time, of course, the rate of formation of ozone also decreases! Thus, they claim that a detailed calculation is needed to determine whether the instantaneous concentration of ozone is sufficient to account for the day-glow. At the present time, the Poona scientists are still carrying out these calculations, and have not yet published their results. One may look forward, however, to a final test of the ozone theory.

THE IONOSPHERE

Free electrons in the earth's ionosphere disappear when sunlight is cut off. We have spoken briefly about the processes that produce and deplete electrons earlier in the chapter. There are three main processes that remove free electrons: *recombination*, in which an electron and a positive ion combine to form a neutral atom or molecule, with the emission of light; *dissociative recombination*, in which an electron and a complex positive ion combine to form two neutral molecules that split off from the original ion; and *attachment*, in which an electron fastens on to a neutral molecule to form a negative ion.

Ionospheric physicists are interested in the relative importance of these processes at different heights in the ionosphere, and have carried out a great variety of eclipse experiments to satisfy their curiosity.

The 1970 eclipse was remarkable because the umbral shadow passed over Wallop's Island in Virginia, which is an established launching site for NASA's rockets. Ionospheric scientists took the opportunity to fire many sounding rockets before, during, and after the totality phase to obtain the distribution in height of the electron concentration. At the same time, they used ground-based ionosondes and other types of equipment to determine the total electron content of a vertical column.

A team from the University of Illinois launched four rockets to an altitude of 170 kilometers. Each rocket was equipped with several Langmuir probes, which are simple devices that pull electrons from the surrounding plasma and measure the concentration of electrons. The rockets were also outfitted with VHF radio receivers. These measured the strength of a polarized radio signal, transmitted from the ground at frequencies of 2.2 and 3.4 MHz. The transmitted signal was linearly polarized, and the receivers detected the rotation of the direction of polarization. This "Faraday rotation" is a measure of the total number of electrons between the receiver and the ground. This quantity was measured continuously during the rise and fall of the rocket, and was later analyzed to give the electron concentration at each height. The series of rockets launched during the eclipse measured the electron concentration at each height as a function of time. These data determined the rate of decay of electrons. The Illinois scientists compared their empirical results with theoretical expressions suitable for either attachment or recombination.

It is no mean trick to launch a rocket at the proper time, speed, and direction to intercept an eclipse shadow which is moving at 1000 miles per hour, and in such a way that the rocket spends half of its trip within the shadow. Nevertheless, the experiment was completely succesful, in that all rockets reached their intended heights, all equipment worked, and in particular, the first two rockets entered the umbral shadow on their initial rise to maximum altitude.

Figure 7.8 shows the electron concentration profile that they measured. Notice that by second contact, the concentration of electrons at all heights between 70 and 100 kilometers decreased markedly, by factors of 2 to 100. The greatest decreases are at the lowest heights. The rates of decrease are very large; that is, it takes only a very short time for the number of electrons to diminish, once the ionizing radiation from the Sun is cut off.

When the Illinois scientists compared the rates of decrease with theoretical expressions, they concluded that below a height of 85 kilometers (that is, in the D region), electrons are lost primarily by an *attachment* process. They could determine the rate at which this pro-

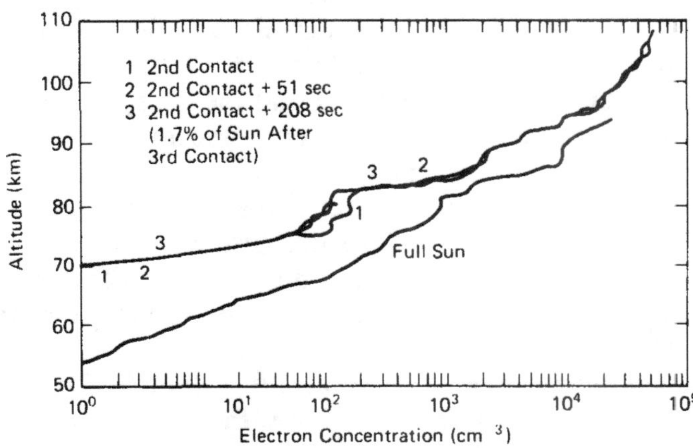

Fig. 7.8. Height distributions of electron density in the D layer of the ionosphere before and during the total eclipse of March 7, 1970. (*Courtesy of E. A. Mechtly et al., "Electron loss coefficients for the D-region of the ionosphere...," in* Journal of Atmospheric and Terrestrial Physics, *34 (1972) by permission of Pergamon Press Limited, Oxford.*)

cess proceeds, but could not identify it among all the possible contenders.

Meanwhile, a group of Canadian scientists carried out almost exactly the same experiment from temporary rocket launch facilities set up at East Quoddy in Nova Scotia on the eclipse track. They launched four rockets before, during, and after totality to a height of about 150 kilometers. The rockets sampled electron concentrations and total electron content during their flight.

Like the Illinois scientists, the Canadians found that the decrease in electron density varies with height; a decrease by a factor of more than 10 was observed at 80 kilometers, changing to a factor of 3 at 95 kilometers, and remaining nearly constant above that height. The Canadians also attributed the loss of electrons to an attachment process that increases the number of negative ions below a height of 80 kilometers, but did not identify the process.

The Canadians also wanted to compare the minimum electron concentrations reached during a total eclipse with those following a normal night. Since they couldn't launch a rocket at East Quoddy during the night, they compared their eclipse values with night-time profiles (electron concentrations at different heights) previously observed at Wallop's Island. The comparison below 85 kilometers was difficult because of night-to-night changes in electron concentrations. However, in the E layer, particularly above 110 kilometers, they found a significant result: the minimum electron densities during the eclipse were greater by more than a factor of 10 than those following an average night. This result suggests that the decay at these greater heights is *not* the result of attachment, but some much slower process, presumably radiative recombination.

Many ionospheric physicists detected changes in the total electron content during the March 7, 1970 eclipse using ground-based equipment. Almost without exception, they measured the Faraday rotation of the linearly polarized radio signal from one or more geostationary, artificial satellites.

As we mentioned earlier, the amount of rotation of the plane of polarization of the radio signal measures the total electron content along the path of the signal. During the eclipse, four artificial satellites (ATS-1, 3, 5, and "Early Bird") were in geostationary orbits at strategic locations. The investigators measured the daily variation of total electron content for several days before and after the eclipse as a control on their eclipse measurements. The total electron content during the eclipse decreased by about 30%.

Depending on his location, each investigator recorded a slightly different time delay between the middle of totality and the minimum in total electron content. The delays ranged between 20 and 40 minutes. The time delay reflects the integrated effect of several mechanisms throughout the ionosphere (such as horizontal and vertical electron transport, or local depletion of electrons), but it hard to interpret quantitatively.

SUMMARY

Ionospheric physicists have concluded from eclipse experiments that the rapid loss of free electrons results from the formation of negative ions in the D layer, from recombination in the E layer, and by transport by ionized winds in the F layer. The higher we go, the slower is the process. In the F layer, transport processes take *hours* to establish a new equilibrium; thus, an eclipse of a few minutes produces virtually no effect, and it takes many hours during a normal night for the electron concentration to decay at these heights.

Ionospheric experiments have been carried out at most of the major eclipses since 1970—particularly, June 30, 1973, October 23, 1976, February 26, 1979, and February 16, 1980. The Indian physicists were particularly interested in exploiting the 1980 eclipse. They mounted over 40 different experiments, mostly from the ground, but some from rockets. The rocket results have not yet been published, and only preliminary results from the ground-based investigations have been published. These reports suggest that the Indian scientists have confirmed, in large part, earlier measurements on the total electron content variations during an eclipse.

SUGGESTIONS FOR FURTHER READING

Books

Goody, R. and Walker, J. *Atmospheres*. Prentice Hall: 1972.

Hargreaves, J. K. *The Upper Atmosphere and Solar-terrestrial Relations*. Van Nostrand Reinhold: 1979.

Ratcliff, J. A. *An Introduction to the Ionosphere and Magnetosphere*. Cambridge University Press: 1972.

8
Waves in the Earth's Atmosphere

During the last decade, meteorologists and ionospheric physicists have been trying to detect waves in the upper atmosphere that are generated during a total solar eclipse. Such waves were predicted in 1970 by two Canadian physicists, G. Chimonas and C. O. Hines. They reasoned that an eclipse causes a sudden cooling of the troposphere and ozonosphere. The rapid chilling of the air in the umbral shadow might create a pressure imbalance that would spread out, in the form of waves, to great distances. Chimonas and Hines suggested that the properties of such waves would test our present understanding of cooling mechanisms and energy transport in the atmosphere. They urged that scientists search for the waves at the forthcoming eclipse of March 7, 1970.

Atmospheric physicists had been familiar with waves in the ionosphere for quite some time. For many years, they had been making continuous records of the "virtual height" of the ionosphere. The virtual height is the height above the ground at which a selected radio frequency is reflected, and it is a measure of the electron density at that height. Because the electron density increases toward greater heights (up to a maximum between 100 km and 300 km), higher frequencies reflect from greater heights. Occasionally these records would show periodic variations; Figure 8.1 is an example. Such a disturbance begins suddenly, lasts for a few hours, and gradually dies away. The records at separated sites look much the same except for a time delay. Put these two characteristics together, and you have the hallmark of a wave-like phenomenon. They are called "Travelling Ionospheric Disturbances," or "TIDs." In 1960, C. Hines identified TIDs as gravity waves.

Two kinds are recognized. The first has periods of 30 minutes or

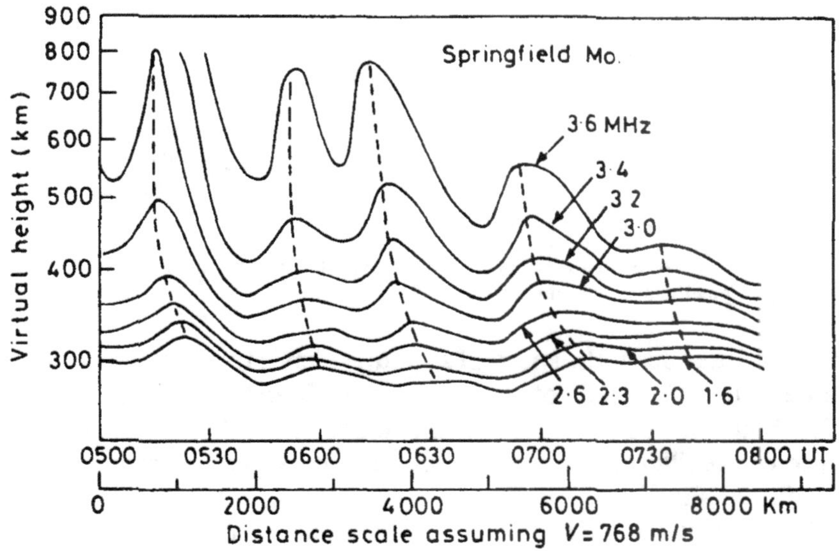

Fig. 8.1. A time record of a travelling ionospheric disturbance over Springfield, Missouri. The "virtual height" is the height above the ground where radio waves of a fixed frequency are reflected from the free electrons in the ionosphere. The wave-like character of the disturbance shows up in the time variations of virtual height. Note that a given maximum of the distrubance propagates downward (see dashed lines). (*Courtesy of J. K. Hargreaves, from T. M. George, "HF Doppler studies of traveling ionospheric disturbances,"* in Journal of Atmospheric and Terrestial Physics, *Vol. 30, no. 211, 1968, by permission of Pergamon Press Limited.*)

longer, wavelengths of a few thousand kilometers, and horizontal speeds of 400–700 meters per second. Aurorae and high latitude geomagnetic activity cause them. The second kind has shorter periods (20–30 minutes), shorter wavelengths (100–200 kilometers), and slower speeds (100–200 meters per second). Thunderstorms cause these.

Let us take a moment to review the properties of atmospheric waves in general, and gravity waves in particular. Any disturbance in the Earth's atmosphere will spread as two main types of waves: sound waves and gravity waves. In a sound wave (see Figure 8.2), an impulse is transferred from one parcel of air to the next by a pressure imbalance. The molecules of air at any particular point flow together to create a pressure excess, and then bounce apart to create a rarefaction. They only oscillate a small distance about their original positions, but transfer energy from one position to the next. The waves travel at a

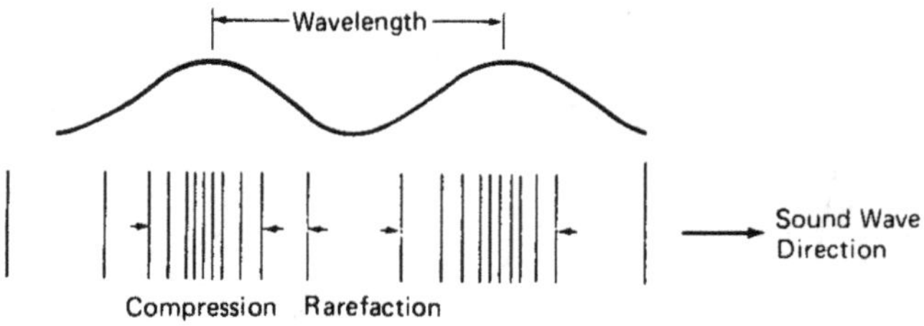

Fig. 8.2. A horizontal sound wave in air. In this "snap-shot," the air is alternately compressed and rarified along the direction of the wave.

characteristic speed, the speed of sound, which is fixed by the local air temperature. The speed of sound is about 350 meters/second, or 780 miles/hour in the D layer of the ionosphere. Notice that each molecule oscillates in the direction the wave is moving, or "longitudinally."

Gravity waves are less familiar to most of us than sound waves. The ripples on the surface of a pond are one example of gravity waves, but a misleading example because they arise only on the air-water *surface,* whereas the gravity waves in the atmosphere are *internal,* and harder to visualize.

Warm air is light and tends to rise, as any balloonist knows. As it rises, it expands and cools. Its upward momentum may carry it above the height where its internal temperature would match the temperature of the surrounding air (see Figure 8.3). It then eventually reaches heights where it is cooler and denser than its surroundings. At that point, it is no longer buoyant and sinks. The process now reverses: the air heats as it compresses in its downward motion, overshoots its equilibrium position and eventually becomes more buoyant than its surroundings. Without friction or loss of heat, the air parcel could continue to oscillate indefinitely in this way.

We've described a relatively slow motion of a particular batch of air, ignoring the reaction of its neighbors. When a batch of air rises, air must flow in from the sides to fill the void. Such flow implies expansion of the neighboring air parcels, and expansion involves cooling and sinking. This secondary motion will create new voids which must be filled in turn. We arrive at the picture shown in Figure 8.4 where a gravity wave travels from lower right to upper left. Notice that the air moves *transversely* to the direction of the wave, unlike the longi-

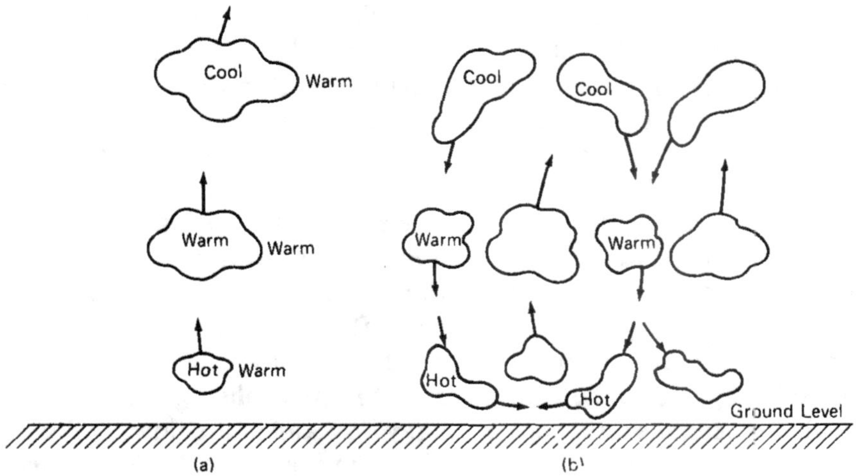

Fig. 8.3. The origin of gravity waves in the Earth's atmosphere. In (*a*), a parcel of heated air rises and overshoots its equilibrium height. In (*b*), cool air sinks and again overshoots its equilibrium height. Buoyancy drives the disturbance.

tudinal motion in a second wave. The transverse motion increases at greater heights because the air density decreases there.

Gravity waves differ from sound waves in several important ways, and these differences aid the experimenter in distinguishing between them. Atmospheric sound waves must have periods shorter (and

Fig. 8.4. A sketch of a gravity wave in the atmosphere. Warm and cool air masses move transversely to the direction of phase propagation, which is obliquely downward in the drawing. Energy flows *upward* and to the left, however.

gravity waves *longer*) than a critical period, which is about 5 minutes. Sound waves travel in all directions with equal ease and at a speed that depends only on the local air temperature. A gravity wave tends to avoid the vertical direction. It can travel no closer to the vertical than a critical angle, which depends on its period; the longer the period, the more nearly vertical the wave can propagate. The horizontal speed of a gravity wave depends upon its wavelength (longer is faster), while its wavelength depends on both its period *and* its direction. Energy flows in the same direction as the phase (i.e. the sequence of oscillations) in a sound wave. In a gravity wave, energy flows at an angle to the direction of phase propagation. The shorter the wavelength of a gravity wave (compared to the scale height in the atmosphere), the larger the angle between the directions of energy flow and phase flow. In Figure 8.4., the energy flows at a right angle to the phase—very different than the situation in a sound wave.

As we mentioned earlier, C. Hines identified travelling ionospheric disturbances as atmospheric gravity waves in 1960. Then in 1970, he and Chimonas predicted that the sudden cooling associated with a total eclipse would generate gravity waves, whose properties would provide a critical check on current theories of energy transport in the atmosphere. In the next section, we examine their theory, and in the following sections, describe subsequent attempts to detect such eclipse-generated waves and to compare them with predictions.

THEORY

When you drop a pebble into a pool of water, circular gravity waves run out from the splash point. If you drop several pebbles, one after the other, spaced along a line, the ripples from each of the splash points will combine at some distance from the line to form a *bow wave* as shown in Figure 8.5. As a ship moves across the sea, it generates a bow wave from the combined action of the individual disturbances it makes at every moment.

Chimonas predicted in 1970 that the umbral shadow of the moon would generate bow waves in the atmosphere in the same way a ship creates a bow wave in the sea. The angle between the wave front and the direction of motion of the shadow is related to the ratio of the speed of the wave to the speed of the shadow, and is typically about 30°.

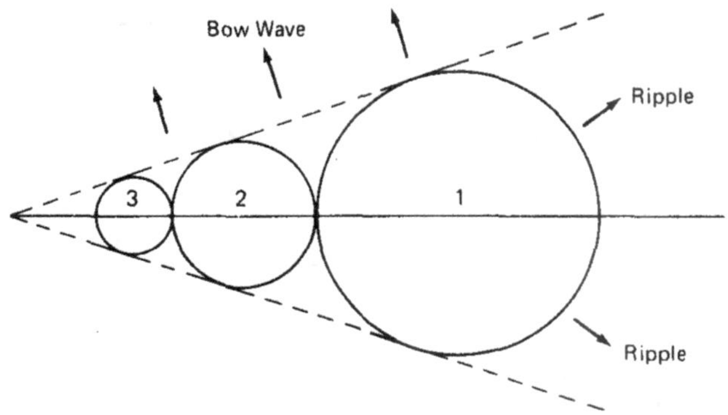

Fig. 8.5. The origin of a bow wave on the surface of a pond. Pebbles were dropped in the sequence 1, 2, 3, and the circular ripples combined to form the inclined bow wave.

Chimonas predicted the period and speed of the waves as seen by an observer located at a great distance from the eclipse path. He also calculated the size of the pressure change that would occur at the ground and in the ionosphere at 200 kilometers. To do this, he examined where the cutoff of solar radiation would have the maximum effect on the atmosphere. The most important region turned out to be the ozonosphere, at about 45 kilometers. Reasonably good rates of heating this layer were available to Chimonas (about 9° per day), and from these he could predict the changes in temperature and pressure that would occur during the passage of the umbral shadow. Figure 8.6, redrawn from his paper, shows the variation of the pressure (in percent) at a height of 200 km above a place that is 10,000 km away from the eclipse track. The pressure change becomes quite large (14%) at 200 kilometers, even though the main cooling effect lies at a height of only 45 km. This is the result of the smaller density of the air at greater heights. Near the ground, the percentage change in pressure is very much smaller, about .001%, but even such changes can be measured with modern microbarographs.

The period of the wave can be calculated by dividing its wavelength, about 10,000 kilometers, by the speed of the shadow, around 650 meters/second. This works out to about 4 hours.

Chimonas and Hines alerted ionospheric physicists and meteorologists to look for such a wave at the forthcoming eclipse of March 7, 1970. They pointed out that the curvature of the eclipse path would focus the northern half of the bow wave on the general area of Califor-

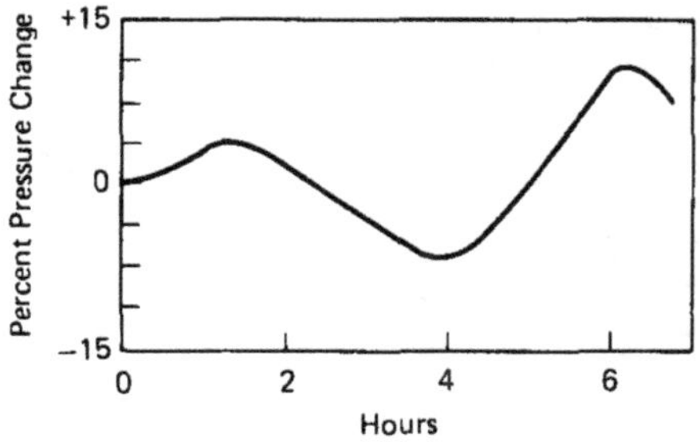

Fig. 8.6. Chimonas' calculated gravity wave disturbance for a place 10,000 km from an eclipse track. Pressure would vary with time, as shown, at a height of 200 km.

nia. Travelling at approximately 320 meters per second, the wave should arrive at California at 1900 hours, Universal Time (UT).

THE MARCH 7, 1970 ECLIPSE

Several experimenters picked up the suggestion of Chimonas and Hines and attempted to observe the predicted waves. They were all very cautious in reporting their results, as experimental scientists generally are. Nevertheless, the Stanford University scientists M. J. Davis and A. V. DaRosa, reported the best evidence in favor of the existence of the waves.

They set up receivers to measure the Faraday rotation (see Glossary) of the polarized radio signals that two satellites in geosynchronous orbit (ATS-1 and ATS-3) transmitted to the ground. Their equipment was placed at Stanford University in California, Clark Lake, California, and Fort Collins, Colorado. From the data, they derived the total electron content of the ionosphere. Figure 8.7 shows the records they obtained at Stanford from the two satellites. A clear undulation began at 1900 UT, the predicted arrival time; however, it had a period of 20 minutes, much shorter than the predicted 4 hours. From the time delays of the undulation among the three sites, they concluded that the wave traveled from east to west as predicted. Traveling Ionospheric Disturbances usually move from north to south, as a result of high-latitude geomagnetic storms. The March 7 eclipse occurred during a lull in a geomagnetic storm that lasted for several days. Thus the direction of the eclipse wave was a point in favor of Chimonas and Hines.

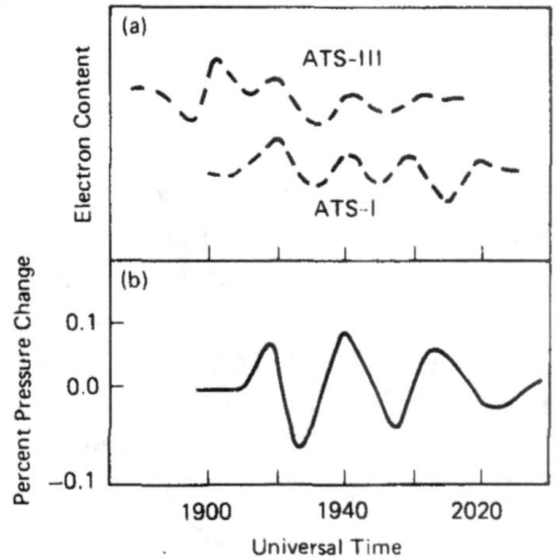

Fig. 8.7. A comparison of the records of Davis and DaRosa, (*a*), with a typical gravity wave calculation by Chimonas and Hines, (*b*). The two agree qualitatively, but with important quantitative differences. (*from George Chimonas, "Atmospheric Gravity Waves induced by a Solar Eclipse, 2," in* Journal of Geophysical Research, *76, October 1, 1971, by permission of the American Geophysical Union.*)

The wave travelled at a speed of 620 meters per second—considerably higher than predicted. The total electron content changed by about 1.5%, but Davis and DaRosa could not relate this change to the predicted pressure change.

After weighing all the facts, Davis and DaRosa concluded:

> The fact that these disturbances of a rather unusual nature occurred near the time predicted for the arrival of atmospheric gravity waves due to the eclipse, and that they came from the direction of the path of the eclipse, implies a possible confirmation of Chimonas and Hines suggestions.

Other observers' results were more ambiguous than those of Davis and DaRosa. G. M. Lerfald and his associates, from the National Oceanographic and Atmospheric Agency and Stanford University, measured the ionosphere's electron density as a function of height from several stations scattered throughout the western United States. They did detect a TID soon after the predicted arrival time of 1900 hours UT. However, the wavefront was arriving from the northeast, rather than the southeast as predicted. Since they had also seen a TID before the eclipse, traveling from north to south, they were reluctant to conclude that they had actually recorded an eclipse-generated wave.

R. Sears, of Lockheed's Palo Alto laboratory, used a different technique from either of the other two groups. He measured the amplitude and relative phase of radio signals transmitted at three frequencies from Fort Collins, CO. Changes in the electron density along the path of such transmissions will introduce a time delay or phase change between signals of different frequency. This is a very sensitive technique for detecting small changes in electron density.

Sears' receiver in California recorded periodic fluctuations of phase, beginning at the magic hour of 1900 UT. The fluctuations had a period of about 25 minutes, and indicated changes in electron density of 1% to 20% in the F-region of the ionosphere. Unfortunately, Sears' experiment was incapable of determining the direction of the TID. Thus, the possibility still remained that the observed TID was not related to the eclipse, but originated as part of the larger, earlier disturbances associated with the geomagnetic storm.

Chimonas and Hines were understandably elated at the positive results found by Davis and DaRosa. The biggest discrepancy between theory and observations at this point was the period, observed at 20 minutes, but predicted at 4 hours. Chimonas and Hines pointed out immediately, however, that Davis and DaRosa were much closer to the eclipse path than they had assumed in their calculations, and should record much shorter periods. They repeated their calculations for an observer closer to the path, and their results are compared with the observations in Figure 8.7. Considering all the assumptions and approximations made in their calculations, Chimonas and Hines had every reason to think that their predictions had been confirmed. However, in view of other conflicting evidence, experimenters resolved to try again at the next convenient eclipse, on June 30, 1973.

THE ECLIPSE OF JUNE 30, 1973

The hopes and expectations of ionospheric physicists for the 1973 eclipse were raised considerably by a set of calculations published by T. Beer and A. N. May in November of 1972. They calculated the trajectories of the wave fronts that the umbral shadow should generate. Beer and May assumed that the waves would travel along the great circle route which begins at the waves' point of origin on the path of totality, and which cuts the initial bow wave at a right angle. This assumption is based on a well-known theorem in wave optics, called "Fermat's Principle of Least Time." They discovered that the waves would focus at three specific locations on the Earth's surface. One of these was in Tibet, where no one was expected to go. Another, unfortunately, fell among the Caroline Islands of the western Pacific, another unlikely eclipse site; but the third fell near the south-western tip of Africa. This focus lay at a great distance from the eclipse path and should have fitted the original theory of Chimonas very well. According to estimates made shortly after the eclipse by A. D. Frost and R. R. Clark (two scientists at the University of New Hampshire) shortly after the eclipse, the intensity of the wave at this focus should have been ten times larger than at the California focus of 1970. Since Davis and DaRosa measured a 1.5% change in the electron content at that focus, a huge effect—as big as 15%—might have been expected in 1973.

By great good fortune the predicted focus was located near the permanent and elaborate facilities of the Max Planck Institute for Astronomy in South West Africa (the point labeled "Schodel" in Figure 8.8). Recognizing their opportunity, the Germans took great pains to try to observe the eclipse wave. They used four different kinds of equipment, including ionosondes, Faraday receivers, and a microbarograph to measure the ground-level changes in pressure. Although a geomagnetic storm was in progress on the day of the eclipse, the Germans got good records.

Beer and May had predicted that the eclipse wave would arrive between 1600 and 1800 hours UT. When the Germans examined their records for this time interval, they saw short period oscillations, but these had begun well before the eclipse and were probably associated with the geomagnetic storm. Similarly, their microbarograph failed to show any eclipse waves at the predicted time.

There was more bad news coming in from East Africa. A group of scientists from Britain, Australia, and Kenya had placed Faraday rotation equipment at four stations strung out nearly perpendicular to the eclipse path (see the circled points in Figure 8.8). They measured variations of the total electron content of the ionosphere for 10 days, including the day of the eclipse. They had hoped to see a change in total electron content as large as 15%, even though they were well

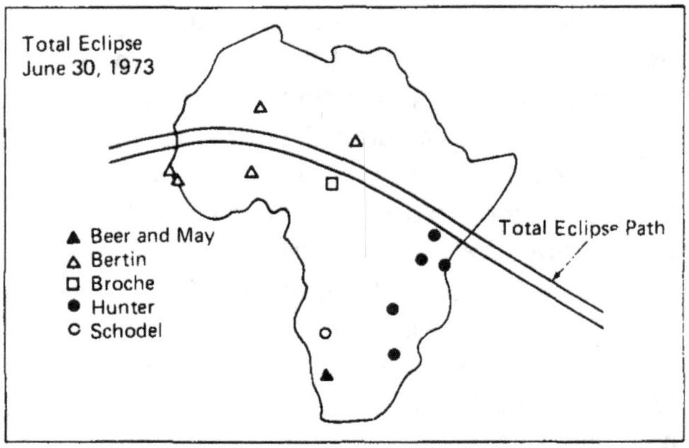

Fig. 8.8. Locations of ionospheric experimenters, relative to the eclipse path of 1973 across Africa. The focal point of gravity waves, calculated by Beer and May, appears as "x".

removed from any wave focus. However, their records showed no changes in the electron content larger than 1% that could be attributed to any eclipse wave.

If these were the only observations taken during the eclipse, one might conclude that Davis and DaRosa's results were a fluke, or that the theorists were wrong—that Chimonas' theory predicted too large an amplitude for the wave, that Beer and May miscalculated where the foci would lie, that Frost and Clark were wrong in predicting a much larger effect in 1973 than in 1970.

However, there is another side to the 1973 story. Two independent experimental groups found confirming evidence for the existence of eclipse-generated waves. P. Broche and M. Crochet, from the University Center of Toulon, set up their experiment at 450 kilometers south of the eclipse path at Fort Archambault, in Chad (see Figure 8.8). They transmitted radio pulses at 8 MHz vertically to the ionosphere, and measured the Doppler shift in frequency caused by the reflection from the F-layer at 300 kilometers. These measurements track the vertical motion of the layer or, equivalently, the changes in the electron density at a fixed height. The maximum optical eclipse occurred at 1220 UT.

Broche and Crochet found not one but *three* different kinds of oscillations in their eclipse records! The main effect was a large, slow oscillation, with a period of a few hours, which they identified with the electron production-recombination mechanism in the F-layer. A rapid oscillation, with a period around 10 minutes began at 1250 UT, and "is certainly caused by acoustic gravity waves." Finally, oscillations with a period of about 30 minutes began at about 1400 UT, and persisted until about 1600 UT. These were too late, presumably, to originate in the eclipse.

Broche and Crochet compared their observations with Chimonas' theory, taking into account their position very close to the eclipse path. The theory predicted about the right time delay and the correct period, but much too small an amplitude. The Doppler frequency measurements, when converted with a bit of theory to an equivalent change in neutral pressure, suggested a 10% change at 300 kilometers. The theory predicts only .1%. Thus, at first glance, Chimonas' theory was off by a factor of 100! But the sense of the discrepancy, you will notice, is that Chimonas predicted too *small* an effect, whereas the observations we've described previously from southwest Africa and east Africa imply that Chimonas predicted *too large* an effect.

Broche and Crochet pointed out that an observer located within the penumbral shadow will see complicated interference effects from waves that originate on different parts of the eclipse track, and that Chimonas' theory does not properly account for these. Despite the inadequacy of the theory, the fact remains that Broche and Crochet, located close to the track, detected a ten-minute oscillation, whereas observers situated at much greater distances did not.

Another group of observers near the eclipse track had a similar experience. F. Bertin, K. Hughes, and L. Kersley operated a series of stations across Africa during the eclipse, as shown in Figure 8.8. At the two western stations, they observed the total electron content, using Faraday rotation equipment. At four other stations, they used ionospheric sounders to record changes in the electron density of the F-2 layer.

Like Broche and Crochet, they discovered oscillations with three different periods in their records. The two western stations recorded a 35-minute wave, traveling roughly from north to south and with a speed of only 80 meters/second, a value too low to reconcile with an eclipse-generated wave. A 42-minute wave was found at three of the ionosonde stations, and appeared to be traveling toward the southeast. However, since one of the stations was north of the eclipse path and should have seen the wave traveling toward the northeast, this wave must also be discarded. Finally, all four ionosonde stations saw an 18-minute wave, propagating from the eclipse region at a speed of about 275 meters/second and in the correct direction.

Bertin and his colleagues concluded:

> It is thus possible that this wavelength and consequent period are characteristic of eclipse-produced waves, a postulate which would be supported by the observation of 18 minute waves during the 1970 solar eclipse.

These different wave periods are confusing, but you must remember that the bow wave originally contains a mixture of wave periods and that bow waves from different parts of the eclipse track combine either constructively or destructively. A distant observer should expect to see only the longer periods. On the other hand, an observer close to the track should expect to see a mixture of periods, and in particular, the shorter periods. All waves die out as they propagate, of course, and may become undetectable despite the predicted focusing effect.

This picture, presented by Chimonas and Hines in discussing the results of Davis and DaRosa, agrees qualitatively with the experience of most of the stations in Africa during the 1973 eclipse, with the main exception of the stations in east Africa. It *is* surprising that the two stations in east Africa nearest the eclipse path saw no waves.

Once again, the attempts of ionospheric physicists to confirm the existence of eclipse-generated waves produced conflicting evidence. The geomagnetic storm in progress during the eclipse certainly did not help. The elements were working against the meteorologists, as well. R. C. Anderson and D. R. Keefer, from the University of Florida, tried to measure ground-level changes of pressure during the eclipse at a station in the western Sahara, but were wiped out by a sandstorm.

In view of the tantalizing results from the 1973 eclipse, scientists resolved to try once again. Their next opportunity arose in 1976.

THE ECLIPSE OF OCTOBER 23, 1976

The path of totality for this eclipse passed over the southeastern tip of Australia and was concave toward the continent. Because of poor weather prospects, a relatively short totality phase, and the low elevation of the Sun in Australia, the eclipse attracted relatively few optical observers. However, this eclipse turned out beautifully for ionospheric scientists because the geomagnetic field was extraordinarily quiet throughout the week before and during the eclipse.

Three Australian groups took advantage of the eclipse to search for gravity waves. Two of these used ionospheric methods, and the third measured pressure changes on the ground with microbarographs. All the instruments were located either on the path or close to it, and thus were in good position to observe short period waves.

A group of physicists from Latrobe University, led by E. Butcher, obtained positive results, but as we shall see, these results were not consistent with the Chimonas-Hines theory and the Australians invented one of their own. They located a digital ionosonde directly on the path to measure the vertical distribution of electrons in the F-layer. Totality occurred between 1640 and 1645 local time. Shortly thereafter, at 1700 hours, a distinct 30-minute oscillation was evident at all heights and "appears to be associated with the eclipse." Two hours after totality, around 1830, a 15-minute oscillation began. The 30-minute oscillation was first detected at the highest frequencies, i.e. at the greatest heights, and, after a short time delay, at lower heights. This effect implies a *downward* propagation of phase. If you will look at Figure 8.4, you will recognize that this result suggests a gravity wave moving *upward* at some angle to the vertical.

Butcher and his colleagues determined the vertical wavelength of the wave from the time delays between different heights; they found a value of 220 km. Next, using gravity wave theory, they calculated the horizontal wavelength, and this worked out to 380 km. They were struck by the similarity between these numbers and the dimensions of the umbral shadow as it passed their station. Was it possible, they asked, that the period and wavelength of gravity waves excited directly *on* the path depend strongly on the characteristics of the shadow itself?

Being located on the path, they did not expect to see the bow wave phenomena predicted by Chimonas and Hines, and had to look for an alternate explanation. They suggested that the rapid cooling of the ozonosphere at 45 km causes a horizontal collapse of the atmosphere, as though it were flowing into a hole. This disturbance travels at the speed of sound, and mainly across the path. The time it takes a sound wave to ride in from the edges of the umbra determines the period of the ensuing gravity wave. An observer located on the path sees waves from the supersonic shadow that has passed him. The period (P) he observes will be Doppler shifted according to the expression:

$$P_{\text{observed}} = P_{\text{wave}} (V_{\text{sound}} + V_{\text{umbra}}) / V_{\text{sound}}.$$

Using these simple assumptions, and taking into account the actual dimensions of the shadow, the Latrobe scientists calculated the period they should have observed, and for good measure, the period that Bertin and Broche-Crochet should have detected at the 1973 eclipse. The table below (Table 8.1) shows their results.

The index N is the number of wavelengths of sound across the horizontal dimension of the shadow. $N = 2$ accounts for the 18-minute oscillation of Bertin and the 35-minute of Butcher. $N = 4$ accounts for the 10-minute oscillation of Broche and Crochet, and the 15-minute oscillation of Butcher. $N = 1$ might even account for the 42-minute uncertain oscillation reported by Bertin.

So far so good, but why didn't Butcher observe a 20-minute oscillation corresponding to $N = 3$? He frankly didn't know, and could only speculate that for some curious reason, only oscillations with indices 1, 2, 4, and possibly 8 are excited. Now, an index of 8 would correspond to a period of about 7.5 minutes. The Latrobe group pointed out that a wavelike variation of about this period (6 to 7 minutes), was found in a completely independent measurement by Butcher of the apparent "tilt" of the E-layer during the 1976 eclipse!

These positive results of Butcher and his group indicated that Chimonas and Hines were at least correct in suggesting that an eclipse would have effects on the ionosphere. However, the phenomena you observe depend strongly on where you are and how you measure it.

There is no better example of this statement than the experience of F. W. Morton and E. A. Essex, another pair of physicists from

TABLE 8.1 Gravity Wave Periods (Minutes)

	1973		1976	
N	PREDICTED	OBSERVED	PREDICTED	OBSERVED
1	39.0	42?	60	—
2	19.5	18	30	35
3	12.9	—	20	—
4	10.0	10	15	15

Latrobe University. They set up three Faraday receivers in a triangular array on the eclipse path in order to measure the variations in total electron content. When they compared their records for the eclipse day with five control days, they could detect no evidence for eclipse waves whatever! They claim this result was to be expected because the eclipse occurred close to sunset, when the total electron content was small and declining so that the eclipse would have to change the content by a large percentage to be detectable.

G. Goodwin and G. Hobson, from the South Australian Institute of Technology, had better luck. They placed four microbarographs on the south Australian coast in an attempt to detect gravity waves at ground level. One site was directly on the path, and another at a predicted focal point for gravity waves. All four microbarographs recorded an oscillation with a period near 23 minutes. The station directly on the path recorded a nice clean oscillation that began shortly after totality. The predicted location of the focus turned out to be wrong, however, and no unusually large amplitude was observed at that station.

Goodwin and Hobson paired the records of the different sites to determine the speed of the wave. They claim that their data is consistent with the bow wave model if the propagation speed were 310 meters/second. However, although the authors do not make a big point of it, their records show some form of pressure oscillation was in progress at each station *before* an eclipse wave could have arrived. This property of their data weakens their final conclusion.

To summarize, the 1976 eclipse contributed more evidence in favor of the existence of gravity waves, but more confusion on the variation of period with distance. Observers on or near the track had good luck in detecting short period waves, as in the 1973 eclipse. A new idea was advanced to explain the bewildering range of wave periods, and to draw attention to the possible role that the dimensions of the umbral shadow play in exciting the wave.

Between 1976 and 1980, few theorists gave attention to the gravity wave problem. S. Kato and his associates at Kyoto University reformulated the problem, but carried out few calculations of practical value. The experimenters were undoubtedly discouraged by the confusing results obtained at a succession of eclipses. Nevertheless, a large number of Indian scientists attempted to detect gravity waves during the eclipse of February 16, 1980.

THE ECLIPSE OF FEBRUARY 16, 1980

This eclipse occurred during a period of very strong geomagnetic activity. The largest geomagnetic storm seen in three months began on February 15, and was still in progress on the 16th. The circumstances, therefore, were far from ideal for the detection of ionospheric waves.

Six independent experimental groups from Indian institutions measured the variation of the total electron content with Faraday rotation receivers throughout the eclipse. Two of these groups set up networks of four stations, and another, a network of six stations. Some of the stations were on the eclipse track, others, far from it. None of these experiments gave any evidence for a wave-like disturbance triggered by the eclipse. In one or two cases, TIDs were seen on the days either before or after the eclipse, but none during. Most of the experimenters saw a gradual decline and slow recovery of the total electron content, with fluctuations but with no definite attributes of an eclipse-generated wave.

Another five groups of Indian experimenters monitored the strength of long-distance radio transmissions from places as far away as Tashkent and Colombo. None of these experiments yielded clear evidence favoring the existence of an eclipse wave, although several of the groups were eager to point out some sort of fluctuation that might qualify. The best looking record is the hardest one to explain.

This one was obtained by monitoring the signal strength of a microwave link between the cities of Dum Dum and Andul. These signals were not reflected from the ionosphere at all, but were transmitted along the line-of-sight between two antennas at a height of about 100 feet above the ground. A distinct oscillation in signal strength began at first contact, and persisted for about an hour after third contact. The oscillation had a period of about 75 minutes, much longer than any period we've mentioned so far. Although the experimenters conclude that the oscillation was an eclipse effect, they don't offer any explanation of it.

Microbarograph records at Hyderabad, north of the eclipse track, showed clear wave motions with a period of 55 minutes, decreasing to shorter periods later on. The oscillations began at the time of eclipse totality. On the face of it, this report is perhaps the best evidence in favor of an eclipse-generated wave of any of those published after the eclipse. Yet once again, the period is much longer than any reported earlier.

SUMMARY

There is no question in the minds of competent scientists that eclipses should cause some sort of atmospheric disturbance, which should amplify at great heights and propagate away from the eclipse path. However, the only detailed theory for the phenomenon, that of Chimonas and Hines, has not been confirmed. Experimental evidence for or against the existence of a specific kind of disturbance—gravity waves—is far from convincing.

We can only hope that some clever scientists will devise an elegant experiment at a forthcoming eclipse, using a technique to observe the ionosphere that is as sensitive as possible. It may very well be that the most convenient method used to date, the measurement of Faraday rotation to determine electron content, is not sufficiently sensitive except when the ionosphere is unusually quiet.

There is also a great deal of work to be done on the theory of the excitation and propagation of gravity waves. Tantalizing hints are all we have to work with at the moment, and it will take strong-minded and optimistic research workers to proceed beyond the present point.

SUGGESTIONS FOR FURTHER READING

Books

Hargreaves, J. K. *The Upper Atmosphere and Solar-terrestrial Relations.* Van Nostrand Reinhold: 1979.

Kato, S. *Dynamics of the Upper Atmosphere.* Reidel Publishing Company: 1980.

9
A Test of General Relativity

What subject is more esoteric or abstract to the average man than relativity? Its concepts seem far removed from everyday experience, and are defended from the approach of ordinary mortals by thickets of complex mathematics. People generally respect it but do not understand it. Only the famous equation $E = mc^2$, with its practical consequences in the construction of the first atomic bomb, has become part of our popular culture. This situation is unfortunate because relativity stands as one of the major achievements of the human mind and, along with quantum mechanics, one of the two revolutions that swept fundamental physics in this century.

We would drift too far off our main subject if I were to try to explain any more than the scope and some of the conclusions of general relativity, as a background to one of its principal experimental tests. I hope this chapter will pique your interest, however, and that you will dig deeper into the subject in some of the books suggested for further reading.

The motion of objects in our everyday lives is governed by the laws of mechanics formulated by Isaac Newton in the 17th century. When combined with Newton's famous law of gravitation, these laws predict the complicated motion of the planets in a very satisfying manner. Newton's laws embody the concept of absolute time—that is, the idea that time runs on at the same rate for anyone, no matter where he is or how fast he moves. Moreover, the laws are limited in the sense that they do not allow an experimenter to determine how fast he is moving in an absolute sense, that is, relative to some "standard of rest," a hypothetical fixed entity somewhere in the universe. Experimenters using the laws of dynamics can only determine *differences* in speed between different objects or *changes* in speed.

A crisis arose in physics when James Clerk Maxwell predicted in 1873 (and Heinrich Hertz later confirmed) that light is an electromagnetic phenomenon and that it propagates in empty space at a very definite speed: 299,791 km/sec. This result implied a standard of rest ("empty space") to which the speed of light was referred. This concept conflicted with the Newtonian idea that *no* standard of rest exists. A. A. Michelson and E. W. Morley in 1887 tried to determine the speed and direction of the Earth with respect to this hypothetical standard of rest by measuring the velocity of light in several directions within their laboratory. They expected to find that the speed of light relative to their laboratory would be smaller in the direction of the motion of the Earth through space. As all the world knows, their experiment showed instead that the velocity of light is the same in all directions and, presumably, for all experimenters in all galaxies.

The laws of mechanics and the constancy of the speed of light were in direct conflict. Enter the special theory of relativity, in 1905. Albert Einstein re-examined our traditional ideas of length and time. He showed that two experimenters watching the motion of a third body will disagree on their measurements of its motion. The readings they get of its speed and direction depend upon the relative motion of the two experimenters. Einstein showed how their measurements could be related ("transformed") so as to agree. In this transformation, not only the space measurements but the *time* measurements must be treated. In other words, time is not absolute and the same for all experimenters, but runs at different rates according to the relative motion of different clocks.

With these radical revisions of our ideas of the basic measurements in physics, Einstein was able to reconcile electromagnetic theory and Newtonian mechanics.

Newton's laws are not wrong. They are indeed, very good approximations of the truth, and work beautifully when the speeds of the objects considered are much smaller than the speed of light. In our daily lives, we never run across a situation where the differences between Einstein's and Newton's theories become apparent. Only the astronomer and the physicist concerned with subnuclear particles need to worry.

The Special Theory relates the observations of experimenters whose relative speed and direction of motion are fixed. What about the measurements of observers whose relative motion is accelerating?

This question led Einstein into a deep re-examination of Newton's ideas on gravitation. In Newton's formulation, the gravitational force of attraction between two bodies decreases as the inverse square of the distance between them. To find their separation, an experimenter must measure their positions *at the same time*. But, according to the Special Theory, time is now relative to the observer. Moreover, according to Newton, the force of mutual attraction readjusts *instantaneously* as the separation of the two bodies changes with their motion.

This concept of an instantaneous change in the force of gravitation conflicts with the ideas of electromagnetic theory. The forces between charged particles also depend upon their mutual separation (and their motion), but they do not change instantaneously. Instead, the changes propagate at the speed of light. Einstein struggled until 1916 to describe the force of gravitation in a manner analogous to Maxwell's description of electromagnetic forces. In Einstein's theory of general relativity, both kinds of forces are described by "fields"—modifications of the properties of empty space by the presence of a body—according to definite physical laws.

Some of the consequences of the general theory are extremely bizarre. For example, a body at rest possesses a certain mass, which is a measure of its resistance to changes in its motion. The general theory tells us that as a body accelerates, its mass changes until, at the speed of light, its mass becomes infinite. As a result, no material object can move at the speed of light. Another bizarre example of the theory was the famous equivalence of mass and energy, $E = mc^2$.

Now photons are pure bits of energy and, of course, move at the speed of light. Einstein suggested in 1911 that a photon with a specific energy (determined by its frequency, or wavelength) possesses an equivalent mass. Thus, like material particles, photons should be deflected by the gravitational attraction of large bodies. In Einstein's own words,

> For it follows from the theory here to be brought forward that rays of light passing close to the Sun are deflected by its gravitational field, so that the angular distance between the Sun and a fixed star appearing near to it is apparently increased by nearly a second of arc. As the fixed stars in the parts of the sky near the Sun are visible during total eclipses of the Sun, this consequence of the theory may

be compared with experience. It would be a most desirable thing if astronomers would take up the question here raised.

Figure 9.1 illustrates Einstein's modest suggestion. During a total eclipse of the Sun, a ray of light from a distant star, *S*, bends near the Sun. To an eclipse observer, *O*, the star seems shifted *away* from the Sun in the direction *S'*. According to the theory, the maximum deflection for a star just grazing the Sun would be 1.75 arcseconds. The deflection should decrease in inverse proportion to the distance of the star from the center of the Sun.

To test Einstein's General Theory an astronomer needs to photograph as many stars as possible in the vicinity of the Sun during a total eclipse. Six months later, when the Sun is no longer in that part of the sky, the astronomer should re-photograph the star field. If the theory is right, the stars in the eclipse photo will appear displaced with respect to those in the uneclipsed photo, and the displacements should vary with distance according to Einstein's prescription. Although this method is simple in principle, it involves tremendous practical difficulties, as we shall see in the remainder of this chapter.

Incidentally, the idea that the Sun's gravitational field could bend light rays was proposed, unknown to Einstein, in 1804 by Johann Soldner, who later became director of the Munich Observatory. Soldner suggested this possibility at a time when the wave nature of light had not been discovered and when Newton's concept of light as a shower of "corpuscles" still explained the known facts. Soldner calculated the deflection of light corpuscles near the Sun's edge, using Newtonian mechanics, and found an answer of 0.875 arcseconds.

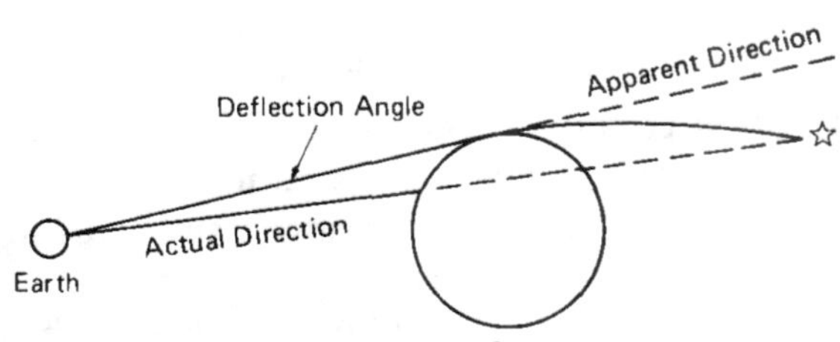

Fig. 9.1. At this point in an eclipse, a star is occulted by the Sun along its actual or true direction to Earth. Because the Sun bends starlight, however, an observer on Earth sees the star exactly on the solar limb. The apparent and true directions define a deflection angle. Einstein predicted that this angle would equal exactly 1.75 arcseconds.

Einstein's result, a century later, was exactly twice as large and arose from very different physical principles.

THE FIRST CONFIRMATION: THE ECLIPSE OF MAY 29, 1919

After Einstein first announced that the sun might bend starlight, E. Freundlich of the Potsdam Observatory examined old eclipse plates to see whether the effect had already been recorded. Unfortunately, no suitable comparison plates had been taken with these eclipse photographs, and Freundlich could not settle the matter.

By 1918, Einstein's General Theory had been confirmed by the analysis of the motion of the perihelion of Mercury, which exceeds the Newtonian value by 43 arcseconds/century. On the other hand, the General Theory predicted a red shift of the Fraunhofer lines on the Sun that had not been confirmed by C. E. St. John. The General Theory was still in doubt, and needed confirmation by the eclipse method. Astronomers around the world resolved to test the theory by measuring the deflection of starlight at the next suitable eclipse.

The Lick Observatory therefore sent an expedition to Washington State to observe the eclipse of June 8, 1918. They erected a huge camera, 40 feet long, pointed directly at the Sun. The eclipse was partly cloudy, unfortunately, and the Lick observers recorded too few stars for an accurate result. The British, therefore, decided to try in 1919.

They organized two expeditions. The first would take the lens of the Greenwich astrographic telescope and another lens from the Royal Irish Academy to Sobral, in north Brazil; the second would take the lens of the Oxford astrographic telescope to the island of Principe, which lies about 120 miles off the African coast in the Gulf of Guinea. Young Arthur Eddington (later to become one of the foremost astrophysicists of the twentieth century) made his debut as an observational astronomer on the Principe expedition. The government grant committee was persuaded to yield the princely sums of 100 pounds for instruments and 1000 pounds for the expeditions.

The British arrived in Sobral in May, which is normally the last month of the rainy season. They settled into a house loaned by the local authorities. Just in front of the house lay the race course of the Jockey Club which "was provided with a covered grandstand, which we found most convenient for unpacking and storage and in the preparatory work." They were greeted by a few afternoon showers, each ushered in by a violent gust of wind.

The morning of the eclipse day was more cloudy than usual; in fact, the proportion of cloud was estimated at 90% at the time of first contact. Fortunately, as totality approached, the sky gradually cleared.

The region around the Sun remained clear except for an interval of about a minute during the middle of totality, when it was veiled by thin cloud. The astronomers took 19 plates with the astrographic telescope, with exposures of 5 and 10 seconds, and 8 exposures with the 4-inch camera. The plates were developed in convenient batches during the night. The party then remained in Brazil until the middle of July in order to take comparison plates. Finally they packed up and went home.

The star images on the plates taken with the astrographic telescope were, unfortunately, out of focus. Apparently the steel tube that held the astrographic lens cooled sufficiently during the eclipse to spoil the focus. The astronomers were especially disappointed because the focus was good on the night of May 27, and the change was only temporary. Without any change in adjustments, the instrument perversely returned to focus when the comparison plates were taken in July. As many as 12 stars appeared on some of the plates. They were all measured and the data were included in the final calculations, but with less weight because of the poor image quality. The 4-inch telescope, however, saved the day for the Sobral expedition. Although only 7 stars appeared on its plates, they were all in sharp focus.

The micrometer at the Royal Greenwich Observatory was too small to handle the 8 × 10 inch eclipse plates. The British astronomers, therefore, worked out the following clever scheme to intercompare the eclipse plates with the comparison plates. They photographed the field of stars in July with the photographic plate placed *backwards* in the camera, that is, with the *glass,* rather than the emulsion, facing the sky. This plate was then placed in contact with the eclipse plate in the measuring machine. It served as a scale, providing undeflected stellar images as references for the deflected images.

Seven stars were measured on each of seven plates. Although the bright corona would mask any star directly at the limb of the Sun, the astronomers could calculate, from the trend of displacement versus distance, the displacement of a hypothetical star at the limb. They found a value of 1.98 arcseconds, compared with Einstein's prediction of 1.75. They estimated that the *random* errors of their final results were no larger than 6%, but as we shall see, it is the systematic errors that matter in this experiment. Nevertheless, the Sobral result was a striking confirmation of Einstein's prediction.

Arthur Eddington and his companion, Mr. Cottingham, had rather worse luck on the island of Principe. Eddington wrote:

> The climate is very moist but not unhealthy. The vegetation is very luxurient and the scenery is extremely beautiful. We arrived near the end of the rainy season, but the gravana—a dry wind—set in about May10th, and from then onwards, no rain fell except on the morning of the eclipse. Indeed, there was a very heavy thunderstorm for an hour and a half before the eclipse, a remarkable occurrence at that time of year. The Sun then appeared for a few minutes, but the clouds gathered again.

Drifting clouds covered the sky throughout the 5-minute totality, but the stoic British carried out their program of exposures right on schedule, and 16 plates were obtained. A few star images showed up on the later plates. To further confound our heroes, the Sun broke through in a perfectly clear sky only a few minutes after totality. Eddington and Cottingham did not remain long enough at Principe to take comparison photographs of the eclipse field. Instead, they photographed the field around Arcturus and left Principe on June 12.

The Arcturus field served as an intermediary between the eclipse plates and a series taken in Oxford in January of that year. The intermediary plates were reversed from those taken at Oxford, and were placed in film to film contact for measurement. Only 4 or 5 stars appeared on about 5 plates. After measurement and reduction of the data, Eddington found deflections at the Sun's limb, ranging between 1.55 and 1.94 arcseconds for a mean of 1.61 arcseconds and a mean error of .3 arcseconds.

Figure 9.2 summarizes the results of the two expeditions. It shows the radial deflection of stars as a function of distance, the predictions of Newton's law of gravitation (the dotted line), Einstein's general theory of relativity (the heavy line), and finally the eclipse results (the light line). The British concluded:

> The results of the expeditions to Sobral and Principe can leave little doubt that a deflection of light takes place in the neighborhood

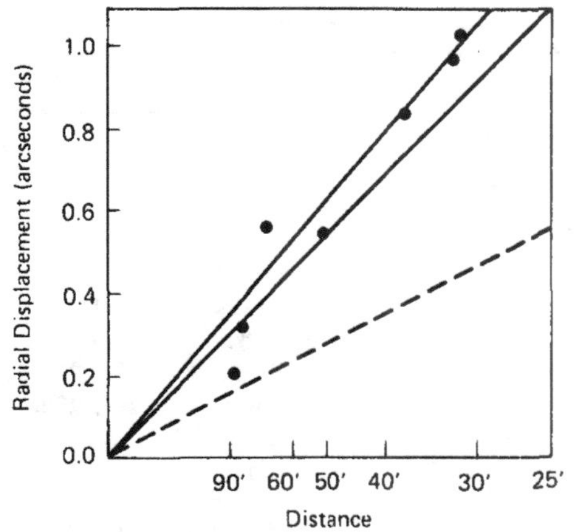

Fig. 9.2. Einstein's predictions of the deflection of starlight, at different distances from the Sun, compared with the British observations of 1919. (*Courtesy of the Royal Society*—Philosophical Transactions of the Royal Society *Series A volume no. 220, p. 291.*)

> of the Sun, and that it is of the amount demanded by Einstein's generalized theory of relativity as attributable to the Sun's gravitational field, but the observation is of such interest that it will probably be considered desireable to repeat it at future eclipses.

How right they were!

FURTHER DEVELOPMENTS

Following the successful expedition of the British in 1919, a whole series of groups attempted to measure the deflection of starlight. Table 9.1 summarizes their results.

Three independent groups observed the eclipse of September 21, 1922. All three found results that tended to confirm Einstein's theory, and that agreed within the errors of measurement. The Lick Observatory expedition managed to photograph more stars at greater distances than anyone else. However, two separate reductions of their eclipse plates gave a slightly different mean value (1.72 as opposed to 1.82 arcseconds) for the deflection of the starlight that just grazes the Sun.

The most discordant note, however, was sounded by the Potsdam group, which observed the May 9, 1929 eclipse in Sumatra. They obtained a much larger result (2.24 arcseconds) and their small mean er-

Table 9.1

Eclipse	Site	θ_0(Arcsec)
May 29, 1919	Sobral	1.98 ± 0.16
	Principe	1.61 ± 0.40
September 21, 1922	Australia	1.77 ± 0.40
	Australia	1.42 to 2.16
	Australia	1.72 ± 0.15
	Australia	1.82 ± 0.20
May 9, 1929	Sumatra	2.24 ± 0.10
June 19, 1936	U.S.S.R.	2.73 ± 0.31
	Japan	1.28 to 2.13
May 20, 1947	Brazil	2.01 ± 0.27
February 25, 1952	Sudan	1.70 ± 0.10

Source: H. von Kluber, "The Determination of Einstein's Light Deflection in the Gravitational Field of the Sun," in *Vistas in Astronomy*, edited by A. Beer, vol. 3, 1960, by permission of Pergamon Press Limited, and in *Gravitation: An Introduction to Current Research*, ed. by L. Witten, by permission of John Wiley & Sons, Inc., 1962.

ror strongly suggested a real discrepancy between theory and observation. Puzzled, the Potsdam group re-examined the British results of 1919.

They learned that the conclusions you draw from the British data depend very strongly on how you reduce the measurements. Moreover, there is always the question of systematic errors: small changes in focal length, the creep of the photographic emulsion, optical or atmospheric effects, and the like, that change the scale of the photographs without disturbing the precision of an individual measurement. The Potsdam group found that omitting individual stars (those, for example, with particularly bad images) sometimes changed the British results significantly. The Germans carefully reexamined all previous measurements and decided that in each case the systematic effects would tend to *increase* the values of the deflections that were previously reported, vindicating their own high value.

The controversy continued to boil during the 1930s. A Soviet group also obtained a large value (2.73 arcseconds) at the June 19, 1936 eclipse. This value is almost 50% larger than predicted! On the other hand, the Japanese at the same eclipse with many fewer stars, found much smaller values that fell *between* the predictions of the Newtonian and Einsteinian theories of gravitation.

The astronomers kept trying. Van Biesbroeck took the same equipment to two eclipses, one in 1947, the other in 1952. If the mean errors he quotes are believable, his two results are significantly different. The Russian astronomer, Mikhailov, agreed. He reduced van Biesbroeck's measurements independently and found 2.20 and 1.43 arcseconds for the 1947 and 1952 eclipses. The Germans persisted. F. Schmeidler obtained 2.17 arcseconds at the October 2, 1959 eclipse in the Sahara. He tried again in 1961 in Italy, but obtained poor results because of the low elevation of the Sun, and once again in 1963, when heavy clouds blotted out the star images.

By the early 1960s, astronomers were convinced that the Sun deflects starlight and that Einstein's predictions are closer to the observations than Newton's. There remained the possibility that Einstein's theory might have to be modified, but the data were still too uncertain to show whether this was necessary.

At this point, a new theory of gravitation appeared: the Brans-Dicke theory. In essence, this theory suggested that the local value of gravitational attraction between two bodies is coupled to the mass-density of the universe. A new quantity—the "scalar field" appears in their equations for the gravitational field. It would take us much too far afield to examine this theory in any greater detail. Suffice it to say that the Brans-Dicke theory also predicts a deflection of starlight by the Sun, but one that is about 8% less than Einstein's.

Here was a real challenge to the astronomers: the smallest mean error of measurement quoted so far was about.1 arcseconds, or 6% of Einstein's deflection of 1.75 arcseconds. To distinguish between the Einstein and Brans-Dicke theory, the error of measurement would have to be reduced by at least a factor of two.

In 1971, scientists at the Departments of Astronomy and Physics at the University of Texas at Austin and the Department of Physics at Princeton University began to plan an expedition to observe the total eclipse of June 30, 1973. They resolved to take every possible precaution to achieve the accuracy necessary to distinguish between the two competing theories of general relativity and to settle the matter once and for all.

THE 1973 ECLIPSE

The eclipse of June 30, 1973 had the second longest duration of totality of any in the twentieth century, and occurred against a rich background of stars in the Milky Way. Thus it offered the Texas-Princeton team a real chance to record more stars more accurately than ever before.

The team selected the oasis of Chinguetti, in Mauritania in West Africa, for their site. This location had several advantages. The length of totality would be 6 minutes and 18 seconds—perhaps not the maximum, but still very respectable. Moreover, Chinguetti lies far enough away from the Atlantic Ocean and high enough in altitude to avoid desert haze. The Sun would appear at the comfortable altitude of 60° above the horizon. In Kenya, the second possible site, totality was shorter and the Sun was lower in the sky. Finally, Chinguetti is a well-ordered community, where equipment could be left safely for 6 months, until the team returned to take comparison photographs.

The team recognized that the key to accurate measurement is precise temperature control of all elements of the telescope and its surroundings. There were three ways to attain this objective: (1) The eclipse and comparison photographs should be taken at the *same* temperature; (2) the equipment had to be *insensitive* to temperature, or (3) all changes in the temperature would have to be *monitored* and accounted for in the reduction of the data. The team used all three of these methods.

First, they built an insulated hut on the site, with double door air-locks (see Figure 9.3). An air conditioning system produced a positive pressure in the interior so that dust could not enter when the doors were closed. They borrowed a fine 9-inch lens from the Paris Observatory for the experiment. This objective produced sharp stellar images over a 6 degree field, but a temperature change as small as 10° Fahrenheit would shift its focus by 3 mm, much too large to be acceptable. Therefore, the telescope tube was wrapped with fiberglass and thermistor temperature sensors were embedded at crucial positions over the whole structure to permit the team to monitor small changes in temperature throughout the experiment.

The emulsion on photographic plates has a tendency to creep differentially, depending upon its exposure and development. Thus, the parts of the film exposed by the bright corona could creep more than the darker parts further from the Sun. This differential creepage could produce an effect very similar but *opposite* to the deflection of starlight the team wanted to measure. To guard against this effect,

Fig. 9.3. The telescope hut at Chinguetti, Mauritania, erected by the Texas-Princeton team for the 1973 eclipse test of general relativity. (*Courtesy of the Texas Mauritanian Eclipse Team and* Sky and Telescope.)

they preexposed their eclipse plates with a very precise grid of 850 artificial stars, which served as a reference for the measurements. Moreoever, they arranged a rotating shutter to dim the bright corona in the center of the field. Finally, they planned to double-expose each eclipse plate: first on the Sun, and then on the star field 10° south of the Sun. The second exposure would show up defects of the lens and errors due to refraction.

The astronomers took as much care with their darkroom procedures as with their actual eclipse experiment. Their darkroom had a separate air conditioner and temperature control. They planned to maintain the darkroom and all the solutions at 68°F, independent of the temperature outdoors or in the observing room. The hut was sealed and maintained at the expected outdoor temperature (85°F) for two days in order to allow everything to come to equilibrium.

A strong north wind had blown for several days before the eclipse, driving the sand before it and filling the air with dust. On the eclipse morning, the wind was still blowing. However, a few minutes before totality and just before the hatch in the hut was opened, the wind almost died and the dust began to thin out. The team went into action according to plan. They double-exposed three eclipse plates: 60 seconds on the Sun, and 30 seconds on the star field 10° south of the Sun.

Because the air was filled with dust, the sky around the Sun was unexpectedly bright, and only Venus was visible to the eye. The team had expected to record over 1000 star images, down to a photographic magnitude of 10.5, but the dust in the air transmitted only 18% of the light. Nevertheless, 150 measurable images, down to a photographic magnitude of 8.5 were found on the plates. Although the outside temperature was 96°F instead of the expected 85°F, the focus of the telescope held beautifully, and the star images were sharp.

The team mothballed the hut, cemented the stops on the telescope to prevent any motion, and went home. Another crew returned early in November to expose the comparison plates. They found everything in order, took the plates, dismantled the equipment, and returned to Austin.

The plates were ultimately delivered to the Royal Greenwich Observatory for measurement on their comparator. Then they were returned to Austin for further measurement on a microdensitometer. Corrections were applied for atmospheric refraction, the motion of the Earth, and the motions of the stars. The shift of the grid points was measured carefully to allow for emulsion creep. Finally, all the measurements were analyzed with an elaborate computer program.

The final result gave the deflection of starlight, extrapolated to the edge of the Sun, as 1.66 ± .18 arcseconds. Figure 9.4 shows how the deflection varies away from the center of the Sun for the 40 stars measured. As you can see, the scatter is fairly large at large distances, and the position of the curve depends strongly on one or two points close to the Sun.

The Texas-Princeton result confirmed once again that Einstein's prediction is far more accurate than Newton's. The result was not, unfortunately, accurate enough to distinguish between the Brans-Dicke theory and the Einstein theory. The Texans gave this experiment their very best, and it was only the occurence of an unpredictable and unavoidable sandstorm that limited the value of their results.

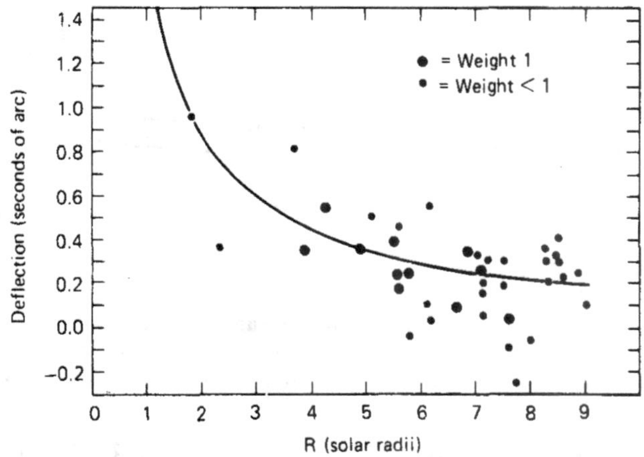

Fig. 9.4. Star deflections versus distance from the Sun, photographed during the 1973 eclipse. (*Courtesy of the Texas Mauritanian Eclipse Team and* The Astronomical Journal.)

THE RADIOASTRONOMERS SPEAK UP

Radioastronomers realized in the early 1960s that microwave radiation from astronomical sources should be deflected from the Sun in the same way as visible light from stars. They were beginning to build interferometers with very long baselines that enabled them to determine the direction of a radio source with a very high angular precision. Thus, the stage was set for an independent, non-eclipse test of the deflection of radiation by the Sun.

The Sun passes several strong radio sources on its apparent path along the ecliptic during the year. Figure 9.5 shows how the apparent position of such a source will shift as the Sun moves past it. You will notice that the source apparently shifts first to the east and then to the west, and this change of direction is a clear signature of the bending of the radio waves.

Radioastronomers encounter several problems in making this test of general relativity that optical astronomers do not. For example, the corona refracts (i.e. bends) microwaves in the same direction as the general relativity effect. Fortunately, the diffraction decreases at larger radio frequencies, whereas the relativity effect is independent of frequency. Thus, by measuring the position of the source simultaneously in two frequencies, an astronomer can distinguish between the refraction effect and the relativity effect. The corona also

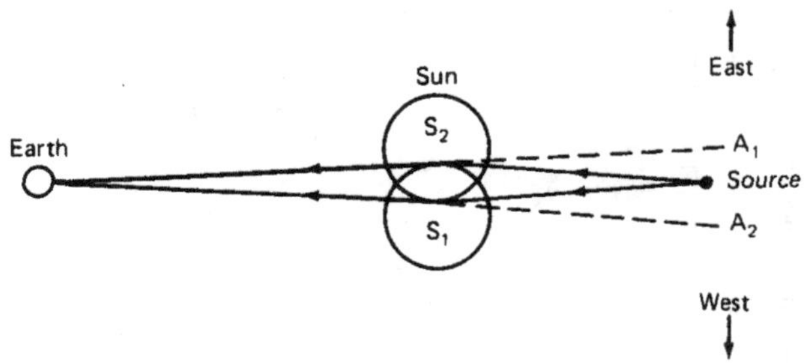

Fig. 9.5. Deflection of radio waves by the Sun. The Sun, at S_1, occults the radio source, but a ray bends around the Sun to reach Earth. The source then appears in the direction A_1, east of its true direction. Similarly, when the Sun moves to S_2, the source appears to shift westward.

magnifies the apparent size of the source by scattering the incident microwaves. This effect limits the distance to which the source can be tracked near the Sun. Despite these problems, the radio method offers the prospect of very high angular precision.

Beginning in 1969, radioastronomers began to use the interferometers available to them to measure the positions of two quasars (3C273 and 3C279) that are close together in the sky and pass very near the Sun. By measuring the change in the *separation* of the two sources instead of the change in the *direction* of one source, the astronomers could eliminate certain spurious effects produced by the instrument and by fluctuations in the Earth's atmosphere.

Between 1969 and 1973, nine different measurements were made of the deflection of these two radio sources. They all confirmed the predictions of Einstein's theory of general relativity. The agreement between observations and theory gradually improved with the precision of the observations. By 1972, theory and observations agreed to about 3%.

In 1974 and 1975, A. B. Fomalont and R.A. Sramek used a new interferometer at the National Radio Astronomical Observatory with a 35 km baseline to test Einstein's predictions more rigorously than ever before. They took advantage of the high sensitivity of this interferometer to measure the positions of three sources that were ten times fainter than the earlier favorites, 3C273 and 3C279. These three sources, moreover, lay in a straight line in the sky.

The outer two were used to calculate the fluctuations in phase of the central source, as all three microwave beams passed through the Earth's atmosphere. The astronomers measured the positions of the sources every 7 minutes for 10 hours a day, and for 13 days in a month. The central source was observed as close as 1.2° from the Sun.

Figure 9.6 compares the measured deflection with the predictions of Einstein's theory. Time is counted from the eclipse of one of the sources by the Sun around April 11. Fomalont and Sramek found a final result for the bending of microwaves at the edge of the Sun of 1.761 ± 0.016 arcseconds. This is 1.007 ± 0.009 times the amount predicted by general relativity. In short, they confirmed Einstein's theory with a precision of less than 1%.

These results were sufficiently accurate to distinguish between Ein-

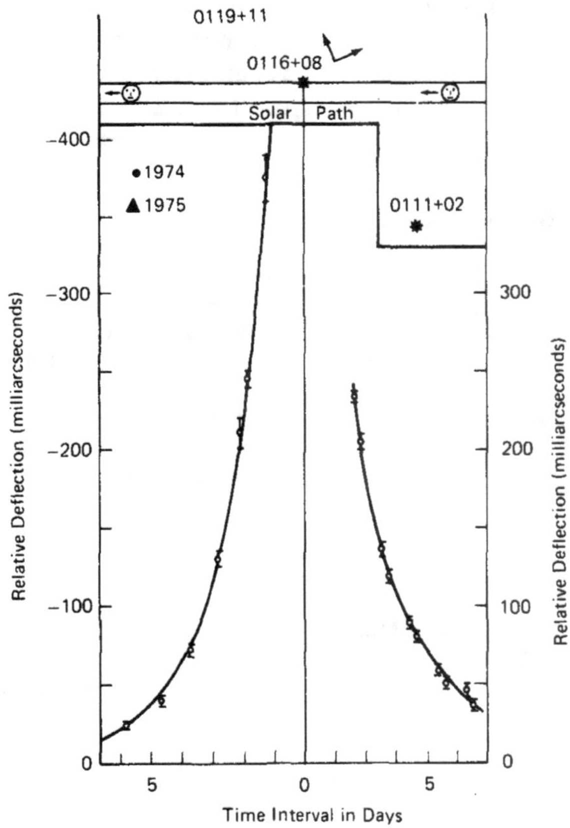

Fig. 9.6. Comparison of predictions of general relativity with observations of deflection of a radio source by the Sun. (*Courtesy of E. Fomalont and R. A. Sramek and* Comments on Astrophysics, *vol. 7, no. 1, p. 29; by permission of Gordon and Breach, Science Publishers, Ltd.*)

stein's theory and the Brans-Dicke theory. The results set limits on the scalar coupling constant that appears in that theory, and with these limits, "the difference between the Brans-Dicke theory and general relativity are slight for most physical applications."

Interferometry techniques with very long baselines have improved tremendously since 1974, and one might hope that the precision of deflection measurements could be increased beyond 1%. However, Fomalont and Sramek expressed their doubts. They felt that the corona sets the basic limit on the accuracy of radio measurements, because even the two frequency technique leaves some uncertainty in the amount of refraction in the corona. Given the past history of radioastronomy, however, I would not be surprised if astronomers came up with some clever new idea. At the moment, the radio method holds the record for precision, and it is unlikely that optical techniques during eclipses will be able to surpass them without some very great change in technique.

SUGGESTIONS FOR FURTHER READING

Books

Einstein, A. *The Meaning of Relativity.* Princeton University Press: 1953.

Sciama, D. W. *The Physical Foundations of General Relativity.* Doubleday and Company: 1969.

Shapiro, I. I. "Experimental Tests of General Relativity," in *General Relativity and Gravitation,* A. Held, editor. Plenum Press: 1980.

Taylor, E. F. and Wheeler, J. A. *Spacetime Physics.* W. H. Freeman and Company: 1966.

Periodicals

DeWitt, B., Matzner, R., and Mikesell, A. "A Relativity Eclipse Experiment Refurbished," in *Sky and Telescope* May 1974, p. 301.

Dyson, F. W., Eddington, A. S., and Davidson, C. "A Determination of the Deflection of Light by the Sun's Gravitational Field," in *Memoirs of the Royal Astronomical Society* **82**, 291, 1920.

10
"I Got Rhythm, You Got Rhythm"

A businessman from Denver arrives in Bangkok after an 18-hour flight. Although it is broad daylight outside his hotel window, his body announces it is ready for its usual 8-hour sleep. Hunger pangs come and go in odd relationship to the clock on the wall. He is experiencing "jet-lag"—the shift of his natural internal rhythms relative to local time.

Each of us (and indeed all plants and animals) is born with an internal biological clock. It governs our daily cycles of sleep and wakefulness, hunger, irritability, and mental alertness. It regulates the daily rhythm of oxygen consumption, body temperature, blood pressure, and blood composition. It controls the female monthly menstrual cycle. It brings on the slow transformations we call "development" or "aging" and decides, ultimately, our life spans.

These internal clocks keep remarkably good time, even without obvious clues from the outside world. Put a man in a cave for six months and he will establish his own pattern of sleep and waking, although his natural day may settle down to anywhere from 20 to 28 hours. Move a crab to a laboratory and it will continue to respond to the daily lunar tides it knew at its native beach.

Some rhythms are locked to subtle external factors and cannot be shifted in time. The daily tides of the atmosphere, and possibly changes in the flux of high-energy solar radiation have been suggested as possible candidates for these external factors.

Most daily rhythms in plants and animals can, however, be shifted—their clocks can be "reset." Our Denver businessman accommodates to Bangkok sunrises within a day or two. A mussel, shifted from one beach to another, modifies its activity in accord with the tides at its new home. Seeds in a laboratory can be made to germinate out of season by varying the length of their artificial daylight.

Each organism has a daily sensitive period (often nighttime) during which its clock can be reset most easily. In these periods, the intensity of light and the temperature of its surroundings affect the ease with which its cycles can be shifted. Thus, some common, cyclical factors, like sunlight, combine with an internal clock to produce the rhythms by which we live.

Biologists want to know how strong the external cyclical factors, such as light, temperature, and humidity, must be in order to reset the internal clocks of animals and plants. Eclipses of the Sun turn out to be a valuable tool in this type of study. An eclipse obviously rapidly changes the light and temperature in ways that differ markedly from a normal sunset or the passage of a large cloud.

Moreover, an eclipse is a "natural" experiment. Animals may be observed in their wild habitats, free of the inhibiting and complicating influences that confinement in a laboratory would incur.

Biologists have drawn different conclusions from observing animals during solar eclipses, both total and partial. Some of their experiments perhaps were not adequately safeguarded with controls. Moreover, the observed changes in animal behavior have not often been related to the stated goal of the experiments, which is an understanding of the mechanisms of setting and resetting internal clocks. Nevertheless, the experimental results are quite interesting.

THE POLISH ECLIPSE EXPERIMENTS

The rising and setting of the Sun, together with the length of the day, influences the sleep cycles of many plants and animals. Perhaps the most obvious example is the behavior of birds, who begin to seek cover during dusk and fall asleep soon after sunset. R. Wojtusiak and Z. Majlert, two Polish zoologists from the Jagiellonian University in Cracow, have observed a variety of birds and insects during seven eclipses of the Sun between 1954 and 1975, in order to determine the least decrease in light (e.g. the shortest partial eclipse) that would produce an observable response.

The eclipse of June 30, 1954 was total in Poland. Wojtusiak and Majlert divided the country into six zones, at increasing distances from the path of totality, and enlisted a corps of volunteers to observe and record the behavior of mammals, birds, bees, and ants. They found that mammals (e.g. cats and dogs) were unaffected by the eclipse even on the totality path. Twenty-eight species of birds were observed, and all showed signs of "anxiety." In particular, larks and sparrows stopped singing on the totality path but continued singing, in increasing proportions, away from the path. Insects reacted more definitely to the eclipse than other types of animals. Bees returned to their hives, en masse, in all the designated eclipse zones, nocturnal moths appeared, and butterflies settled in the grass as though it was true night. These results were confirmed during the eclipse of February 15, 1961 which was partial and reached a magnitude of 87% to 94% in Poland.

The smallest partial eclipse occurred on October 2, 1959, when only 19% of the Sun appeared eclipsed in Cracow. Birds exhibited no significant response but the most sensitive species, bees, returned promptly to their hives. The eclipses of 1966, 1968, 1971 and 1975 added to the experimental data.

The Polish investigators concluded from their extensive study that (1) the response of each species simulates its response to true dusk and dawn, (2) that the strength of the response corresponds directly to the magnitude of the eclipse, and (3) that species differ markedly in their sensitivity to the stimulus. The Poles speculated at length about the electromagnetic radiations or atmospheric factors that might be responsible for the observed behavior, but carried out no further experiments to determine these factors.

As an interesting sideline, the Poles also studied the effect of a total eclipse (in 1954) on the navigation of pigeons. Later investigations have shown that pigeons probably employ several navigational aids: the Sun and the stars (whose changing positions they correct for automatically), the polarization of the daytime sky, the Earth's magnetic field, and possibly an accurate memory of their flight paths. The Poles suspected that pigeons are also sensitive to infra-red and radio waves from the Sun. Since an eclipse affects these different wavelengths by different amounts, they hoped to test their hypothesis during an eclipse.

Sixty-four pigeons were transported from Cracow, divided into four groups, and released at different times. The first, a control group, was released at dawn of the eclipse day, groups two and four, respectively, were set free a half hour before and after the total phase, and group three during totality.

The pigeons released at dawn and before the eclipse returned more rapidly and in greater numbers than those set free during or after totality. Wojtusiak and Majlert assert that these results "strengthen the hypothesis that infra-red and short wave radio waves are sensed," and that their absence tends to disorient pigeons. Considering all the other techniques pigeons use in navigating, however, it is not clear that the Poles had proved their case.

THE INDIAN ECLIPSE OF FEBRUARY 16, 1980

Indian zoologists and biologists carried out a large number of experiments on the effect of a total eclipse on animal and on human behavior. On the whole, their experiments were more varied, more sophisticated, more objective, and better controlled than earlier attempts. To date, only preliminary results have been reported, and the connections between the results and biological rhythms are often tenuous.

Once again, the behavior of birds during an eclipse attracted attention. S. Sengupta, of the Zoological Survey of India, observed two species in the Calcutta Zoo during the eclipse. The Great Whistling Teal normally spends its days dozing by the lake in the zoo, and flies off at dusk to forage for food. One might expect a total eclipse to trigger the flight prematurely. However, for some reason (presumably the overriding influence of the Teal's internal clock) the eclipse had no effect. Java sparrows, on the other hand, are active during the day and sleep after dusk. During totality the sparrows roosted in preparation for sleep, but could not fall asleep! Sengupta concluded that the eclipse stimulus was sufficiently out of phase with the sparrows' rhythm to prevent normal sleep.

Nematodes are roundworms that inhabit freshwater streams and ponds, and brackish or salt water, in enormous numbers. They are tiny—only 16 to 30 thousands of an inch long, and can be seen clearly only under a microscope. Saltwater nematodes live in the sand of beaches, between the low- and high-water marks, and migrate toward the surface at low tide. At most locations, two low tides occur per day, one of them about four to six hours after the Moon crosses the local meridian. One might expect that the tidal movement of seawater over a beach would be the primary external influence regulating the daily activity of these worms. However, it seems that the intensity of sunlight (i.e. the temperature of the beach sand) predominates over the tidal effect. What effect, then, could we expect an eclipse to cause?

A. H. Parulekar and his colleagues at the National Institute of Oceanography at Goa, decided to find out. They sampled the low-tide population of nematodes below the surface, at the high-, mid-, and low-water marks on the Goa beach for two days before, two days after, as well as on the eclipse day.

By coincidence, low tide on eclipse day occurred at about the same time as totality. Since the beach temperature decreased rapidly (and abnormally) during the eclipse, the experimenters could separate the influence of temperature and of tidal movements of seawater. They found that the nematodes *avoided* the surface at low tide in eclipse day, a distinctly abnormal behavior. The experimenters have not published their conclusions yet, but we might anticipate that the sand temperature is the predominant signal that regulates the daily migration cycle of the worms.

The basic source of food for all sea creatures is, ultimately, plankton. It consists of a mixture of microscopic, free-swimming animals (e.g. protozoa, the larvae of shrimp, and jelly fish) and plants (diatoms and blue-green algae). The animals in plankton respond rapidly to changes in the intensity of sunlight by migrating vertically. Some species move down by day and up by night, while others congregate at the surface just after sunset. Will the short, but intense, decrease of sunlight during an eclipse initiate an abnormal migration? Will the abnormal behavior persist? T. S. S. Rao and his colleagues from the National Institute of Oceanography tried to answer these questions.

They collected samples between the surface and a depth of 25 feet in the sea off Goa, on eclipse day and at the same time (3:30 pm) on the days preceding and following the eclipse. As you might guess, the quantity of plankton Rao measured during the eclipse totality was distinctly larger than usual for that time, both at the surface and at depth. This result indicates that even a two-minute eclipse is sufficiently long to activate migration.

R. S. Khan of the Zoological Survey of India carried out a similar experiment on freshwater plankton, at a reservoir on the totality path. He found that the species that normally migrate upward by night moved upward during the eclipse. All species returned to their normal behavior by the morning following the eclipse, however.

You will remember that the Polish investigators learned that mammals are much less sensitive to an eclipse than invertebrates. V. P. Dixit and associates from the Psychobiological Laboratory of the University of Rajastan confirmed this conclusion. They observed a wild troop of Rhesus monkeys during the eclipse and on the preceding and following days. They searched for any significant changes in several types of individual behavior (e.g. self-grooming) and social behavior (e.g. cradling infants). None were found.

The Indian population was enormously excited by the prospect of a total eclipse. Part of the excitement was for religious reasons. Some Hindus hoped to gain extra merit by bathing in certain sacred lakes during the eclipse, and several hundred thousand pilgrims traveled to Kurushetra in the west of India for this purpose. But a great number of people were simply frightened by "old wives tales" about the evil influence of a rare and terrible event like an eclipse. Indian physicians and medical researchers therefore carried out elaborate experiments on human subjects to search for any detectable change in their metabolism.

For example, scientists from the Department of Physiology of the University of Madras and the Department of Biomedical Engineering of the Indian Institute of Technology, in Madras, undertook a series of electrophysiological and neurochemical studies of human subjects. They established two camps on the path of totality. At Hyderabad, a Yoga meditator was connected to an electroencephalograph and his brain waves (alpha, beta, theta, and delta) were recorded. The poor fellow was also rigged to a polygraph in order to measure his respiration, blood flow, heart sounds, and low-level muscular activity. An epileptic patient and a "normal" individual were similarly encumbered. Blood and urine samples were collected before and after the eclipse, and the levels of cholestrol, sugar, urea, proteins, enzymes, dopamine, and other substances were analyzed.

For good measure, these scientists also observed the behavior of several albino rats, and administered intelligence and memory tests to a group of college students before and after the eclipse.

The investigation on human subjects produced completely negative results. Let us hope they will reassure the Indian public. The rats on the other hand, behaved very peculiarly. They were hyperactive before the eclipse, but fell inert during the eclipse. During totality "they were blocking together in one corner of the cage with their eyes closed, looking almost dead." One rat died immediately after the eclipse, while the others gradually recovered. Evidently the investigators were removing the brains of rats throughout the eclipse in order to measure the levels of certain chemicals. It's possible that the experimenters themselves frightened the animals. In any case, we can conclude that humans are unaffected. Thank goodness!

11
Future Experiments: Is the Sun Shrinking?

In earlier chapters, I emphasized that experiments at total eclipses are no longer the only way or even the best way of investigating problems in solar astronomy or in any of the other disciplines we've mentioned. Observations outside of eclipse, especially from space vehicles, have superceded many kinds of experiments, particularly in astronomy, that formerly required an eclipse. Will this trend continue? Will total solar eclipses gradually lose their appeal to professional scientists and only attract enthusiastic amateurs?

Nobody can predict the future, let alone the future course of science, with any confidence. However, I would say the answers to these questions depend on several definite factors. First, there is no doubt that many scientists who have recently carried out eclipse experiments want to confirm or extend their results. As we have seen, some experiments yield ambiguous or controversial results that need confirmation. Among these are the search for iron carbide dust particles, the determination of interplanetary dust orbits, the search for atmospheric gravity waves, and the measurement of temperatures and gas motions in the corona. Other experiments, such as the 1981 airborne infra-red measurements, have yielded exciting results on the first try and should be followed with more detailed studies. Still other experiments need better quality data to strengthen preliminary conclusions. If the group of scientists currently engaged in "eclipse science" can compete successfully for resources, they will surely want to continue their work.

Some astronomical questions are more easily settled by ultraviolet or x-ray observations from a satellite, during a period of months, than with visible light or radio observations during a five-minute eclipse. Thus, given a choice, some scientists would prefer to design experiments for space. But even though it may result in proportionately more or better data, a space experiment may cost 100 to 1000 times more than an eclipse experiment. Thus, we can expect that the number of space experiments that will be funded in the next decade will be very small indeed, and limited to studies with a relatively assured result that cannot be done in any other way.

On the whole I think total eclipses will continue to fill a real need for relatively cheap, exploratory investigations that have special requirements. A great deal depends upon the ingenuity of scientists. Will they continue to dream up clever eclipse experiments to solve interesting and important scientific problems? I can't answer this question definitely, but the evidence at the moment suggests they can and will.

For example, let's consider the Dunham's experiment to measure the diameter of the Sun. The Sun's diameter is expected to change, as the Sun evolves, over periods of billions of years. The interesting question is whether it changes significantly (i.e., measurably) within our lifetimes. Let us take a moment to understand how this question arose in the first place.

A variety of evidence (such as mountain glacier advances, and the composition of the Greenland ice cap) indicates that the Earth's climate has varied considerably during the past 5000 years. There have been periods of abnormal cold such as the "Little Ice Age" in the seventeenth century and periods of unusual warmth, as in this century. Climatologists are searching for explanations for these variations. Their mathematical models of world climate indicate that a change of even a percent or two in the Sun's total radiation would profoundly affect the Earth's climate. In fact if the Sun were to darken by only 4 to 6%, the Earth's oceans would freeze solid, the world's continents would be buried in snow and no subsequent brightening of the Sun could revive the frozen Earth. The change would be irreversible because, once painted white, the Earth would reflect nearly all the incident sunlight.

Thus two questions arise: (1) Has the Sun's luminosity varied, even at the 1% level, during the history of the Earth? (2) If so, can such changes be linked to known climatic changes?

Astronomers have tried to detect changes in the Sun's total radiation, using measurements at ground level or at the tops of high mountains. Charles Abbott, of the Smithsonian Institution, spent 30 years gathering observations and concluded that the Sun has varied during this century. He had to correct his raw data, however, for the effects of absorption and scattering in the Earth's atmosphere in order to estimate the amount of energy that the Sun emits. A recent reanalysis of his data has shown that these corrections cannot be made with an accuracy approaching 1%, and that real changes between 1923 and 1952 are smaller than 0.2%.

One way to avoid the complications introduced by the Earth's atmosphere is, literally, to rise above it—to lift instruments above the atmosphere. Several attempts have been made during the past ten years, with ambiguous or controversial results. For example, two geophysicists from the University of Denver measured the Sun's output in 1968 and again in 1978 from a high altitude balloon. They reported a "real" 0.4% increase over this decade, but doubts have been raised whether their instrument could be calibrated to this accuracy. Similarly, scientists from the Jet Propulsion Laboratory flew an absolute pyroheliometer (a device that measures the flux of solar radiation over the whole spectrum) aboard a rocket, first in June, 1976 and again in November, 1978. They also detected an increase of 0.4% in the solar "constant" and estimated their probable error at about 0.1%. Similar instruments have been placed aboard Nimbus 6 and 7 weather satellites and aboard Solar Maximum Mission. Although interesting short-term fluctuations have been recorded, a permanent change in the solar constant, if any, was too small to detect during the flights of the satellites. A continuous monitoring program from space is needed to settle the question by direct measurement, and such a program is beginning now.

But there are other, more indirect, ways to search for changes in the Sun's luminosity. S. Sofia and his colleagues at the Goddard Space Flight Center pointed out that the Sun should shrink in size if it grows dimmer. Their calculations indicate that a one-percent decrease in the solar luminosity should lead to a 0.72 arcsecond decrease in the solar radius. Since the "measured" changes in the luminosity are, at most, 0.4% and possibly smaller than 0.2%, the change in the radius may be only 0.2 arcseconds over a period of a decade. How can such a small change be detected?

Since 1836, observers at the Royal Greenwich Observatory have measured the position of the Sun each day as it crosses the meridian, in order to determine the correct astronomical time. They observe the

Sun visually, allowing its image to drift across a cross-hair in the eyepiece of their transit instrument, and timing the interval between the passages of the east and west limbs. J. A. Eddy and his associates at the Center for Astrophysics in Cambridge, Massachusetts, examined the visual records extending from 1836 to 1953, in an attempt to extract any solar diameter variations. Despite the bias of individual observers, they detected a long-term decrease of 2.25 arcseconds per century in the horizontal solar diameter and 0.75 arcseconds/century in the vertical diameter. (See Figure 11.1.) Now these are *enormous* rates of decrease—the Sun would vanish to a point in only 1000 years at these rates!

I. I. Shapiro, of MIT, immediately contested these results of Eddy. In 1980, he assembled observations of 23 transits of Mercury between

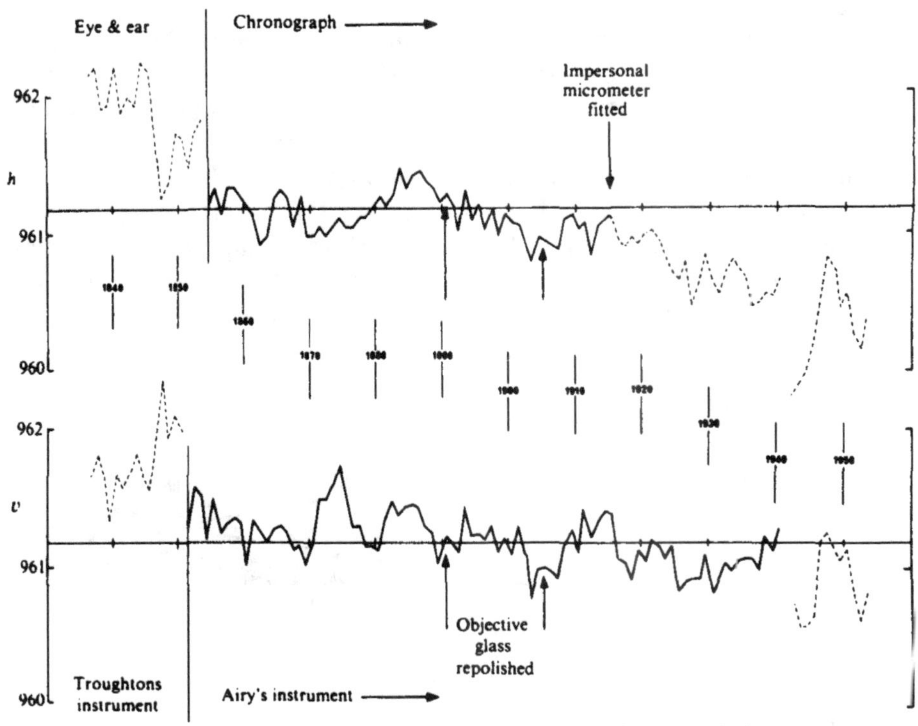

Fig. 11.1. Annual mean values of the Sun's horizontal (*h*) and vertical (*v*) radii, derived from visual meridian circle measurements made at Greenwich. The upper curve shows the century-long decline reported by Eddy and Boornazian. (*Reprinted by permission from* Nature, *vol. 288, p. 548; copyright © 1980 Macmillan Journals Limited.*)

1736 and 1973. Since Mercury's orbit lies closer to the Sun than Earth's, we see the planet cross, or "transit," the face of the Sun about 13 times in a century. The orbits and periods are known well enough to determine the solar diameter from the times of transit and consequently to decide whether the solar diameter is changing between transits. Shapiro found that the solar diameter has not decreased more than 0.3 arcseconds per century.

Meanwhile, the astronomers D. W. and J. B. Dunham had proposed an even more sensitive method to determine the solar diameter. They recognized that the diameter of the umbral shadow at a total eclipse is an indirect measure of the solar diameter, *if* we know the diameter of the Moon and the distances of the Moon and Sun accurately enough (see Figure 11.2). We do, in fact, know the Moon's diameter and the lunar and solar distances can be calculated precisely enough for any given eclipse.

There is one small catch to this clever idea. The Moon, unfortunately, has mountains and therefore its limbs are not perfectly round. The Moon's edge is rough, with irregularities as large as 0.5 arcseconds. Moreover the mountains at the Moon's limb change from one eclipse to another, because the Moon "librates"—it appears to waggle back and forth in both the east-west and north-south directions because its orbit is slightly elliptical. If we attempt to measure the umbral shadow diameter at the *center* of the path of totality, uncertainties in the profile of the Moon's east and west limbs will cause

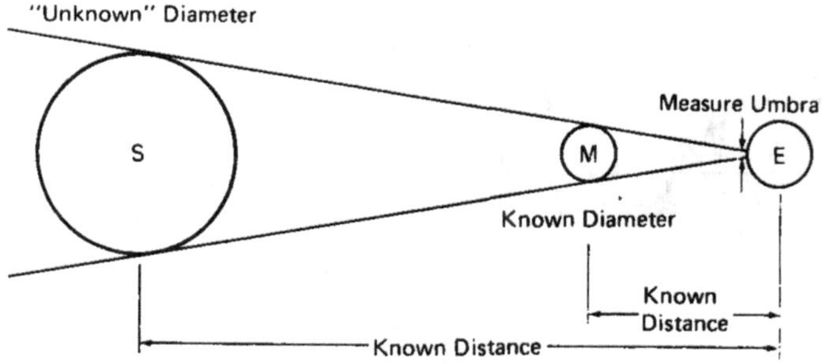

Fig. 11.2. The geometrical basis of the Dunhams' method of determining the solar diameter from the size of the umbral shadow. The known diameter and distance of the Moon, together with the measured diameter of the umbra, define a cone which is tangent to the Sun. Since the Sun's distance from Earth is also known accurately, the cone determines the Sun's diameter, to within 0.1 arcsecond or less.

unacceptable uncertainties in our result. Fortunately, the Moon waggles in the north-south direction by at most one degree. Thus if we measure the umbral diameter in the north-south direction, the uncertainties of the Moon's profile at its poles will have a much smaller effect—that is, we obtain a more accurate result.

The Dunhams' method in practice consists of stationing visual observers across the predicted northern and southern limits of the path of totality. Each observer must decide *only* whether he sees a totality phase, however short, or not. If he stands just inside the limit, he will see the Sun shrink to a point—a Bailey's bead. If he stands just outside, he never sees a bead. This is an easy judgement to make. A line of observers can easily establish the geographical location of the umbral shadow's limit to within a hundred meters. This uncertainty projects to an error in the diameter of the Sun of only .05 arcseconds. Thus a simple "yes-no" visual observation by a group of observers results in high astronomical precision. But whether the precision attainable in this experiment is sufficient to reveal a change in the solar diameter within a "reasonable" time (say, a decade) depends on the actual solar change. Only time will tell.

However, the Dunhams and their collaborators have already obtained a preliminary but significant result, thanks to the far-sighted efforts of Sir Edmond Halley. He anticipated the Dunhams' method back in 1715 and invited amateur eclipse watchers in England to report their impressions during the total eclipse of May 3, 1715. The Dunhams reviewed the evidence that Halley compiled and established the geographic north and south limits of the eclipse path. Applying modern computer codes, they reconstructed the positions of the Sun and Moon and so derived the Sun's vertical diameter for that May morning 267 years ago.

They also observed the eclipses of October 23, 1976 and February 26, 1979, mobilizing teams drawn from the International Occultation Timing Association. Then they applied their method to these contemporary eclipse observations. Table 11.1 summarizes their results.

The method yielded a precision of better than 0.1 arcseconds (or about 70 km on the Sun), which, while admirable, was insufficient to detect any changes between 1976 and 1979. However, by combining Halley's 1715 eclipse and the 1979 eclipse, the Dunhams found a significant decrease: -.34 arcseconds. This is much smaller than the 2.0 arcseconds expected in 264 years from Eddy's rate of contraction, but

Table 11.1 Solar Radius Changes (arcseconds)

1976 to 1979	+.01 ± 0.08
1715 to 1979	−.34 ± 0.2

consistent with the limit of 0.4 arcseconds set by Shapiro. If Sofia's calculations are correct (and D. Gough, of Cambridge University disputes them), a radius decrease of .34 arcseconds would correspond to about a half percent decrease in the solar constant since 1715. We should expect then that the world's climate would have cooled since 1715). In fact it has warmed, so even the *sign* of the Sun's influence may not be understood.

There is clearly much more to be done on this subject. The Dunhams and A. Fiala, from the U. S. Naval Observatory, repeated their experiment at the Indian eclipse of 1980 and plan to continue accumulating data for at least a decade. Annular eclipses may also help, although their observing technique has to be refined. On the theoretical side, a debate rages among specialists on the possible numerical relations between a radius change and a luminosity change, and on the possible internal processes that might cause the Sun to shrink. Good fun and good science await us! If this experiment is typical of the kind scientists will continue to develop, we can expect a long, healthy future for total eclipses.

SUGGESTIONS FOR FURTHER READING

Books

Sofia, S., Dunham, D. W., and Fiala, A. D. "Determinations of Variation of Solar Radius from Eclipse Observations," in *The Ancient Sun* R. O. Pepin, J. A. Eddy, and R. B. Merrill, editors. Pergamon Press: 1980.

Periodicals

Dunham, D. W. and Dunham, J. B. "Observing Total Solar Eclipses from Near the Edge of Predicted Path," in *Moon* **8**, 546, 1973.

Fiala, A. "Observations of a Probable Change in Solar Radius between 1715 and 1979," in *Science* **210**, 1243, 1980.

12
The Great Hawaiian Eclipse

Most total eclipses are best seen from remote, uncomfortable places such as the middle of the Sahara or the mosquito coast of Surabaya, but not the eclipse of July 11, 1991. Its path of totality passed right over the island of Hawaii, which is famous for its hospitality, its white beaches, and its balmy weather. Thousands of ordinary tourits who had never seen an eclipse decided this was the one to see. All through the autumn of 1990, the airlines and hotels planned for a minor invasion. They were not disappointed; the turnout was one of the largest ever seen.

The Hawaiian eclipse was a huge media event. Hundreds of journalists and photographers, a dozen television and radio crews from all over the world descended on the island. For weeks before, nobody talked of anything else. Schoolchildren were equipped with dark slides to view the eclipse and preparations were made to bus them to favorable locations. The local police organized a "war room" to cope with the expected traffic jams. The T-shirt business boomed and a special "eclipse haircut" was invented. Hawaii fell in love with the whole idea.

Amateur eclipse watchers had a difficult choice that year. The eclipse path (see Figure 12.1) would also pass over Baja California, where the weather prospects were equally good and where totality would last 6 minutes and 50 seconds—the longest until the year 2132. On Hawaii totality would last only 4 minutes and 11 seconds. In the end, many eclipse watchers decided in favor of the Baja. Very few eclipse "trekkies" considered watching from Brazil, because of its distance and uncertain weather.

The professional solar astronomers also had a difficult choice. Naturally they wanted the longest possible totality, which argued for the

Fig. 12.1. The path of the total solar eclipse of July 11, 1991.

Baja. But at the 13,700 foot summit of Mauna Kea, one of the three enormous volcanoes that created the island of Hawaii, stands one of the world's largest collections of big nighttime telescopes (Figure 12.2). Ordinarily these telescopes are never pointed at the Sun because they would behave like solar furnaces: after only a few seconds in full sunlight, their secondary optics would evaporate. During a

Fig. 12.2. Some of the major observatories at the summit of Mauna Kea. Photo courtesy of the National Optical Astronomy Observatories.

total eclipse, however, the corona is only as bright as the full Moon, and so the universities that owned these telescopes could be persuaded—reluctantly, sometimes—to lend them to the solar astronomers. Only once before, in 1961, had such an opportunity arisen. The total eclipse of that year passed over the 1.9-meter telescope of the Observatory of Haute Provence.

Solar astronomers wanted a big telescope in order to collect more light during the eclipse, since more light generally means more accurate observations. Solar telescopes usually do not need huge mirrors because the Sun is so bright. The largest (on Kitt Peak, near Tucson) is a mere 1.6 meters in diameter, a pygmy in comparison to the mirrors on Mauna Kea, which range up to 3.6 meters. (The Keck Telescope, with a segmented mirror 10 meters in diameter, was just being finished in 1991.)

Mauna Kea offered additional advantages. Its superb "seeing" (very smooth airflow above the telescopes) would guarantee sharp solar images. Also, the air high above the island is very dry, which would permit eclipse observations at very long infrared wavelengths.

Moreover, Mauna Kea is fully equipped with dormitories and workshops to support large groups of astronomers. All these factors drew over a dozen teams from around the world to Hawaii, despite the shorter duration of totality. Only seven of these were selected to observe at one of the big telescopes. The Japanese and Slovakian teams, who brought their own specialized instruments, decided to go to Baja California.

Mauna Kea does have some drawbacks as a place to make astronomical observations, however. It is a thoroughly extinct volcano, so an eruption is not a serious threat, but at the 13,700-foot summit, the brain slows down due to lack of oxygen. You can add the same column of numbers three times and come up with three different answers. Every step in setting up equipment has to be planned in advance, preferably in the offices at 9,100 feet. Some astronomers have trouble breathing or experience severe headaches at the top of Mauna Kea. But none of these factors weighed very heavily against the once-in-a-lifetime chance to observe an eclipse with a meter-class telescope.

For several years before the eclipse, a number of important scientific questions had been sharpened to the point where specific observations might provide answers. Among these were the problems of how the corona is heated, what its true density and magnetic field within its fine-structures are, and how rapidly does this fine-structure change shape. Earlier eclipses had raised questions about the structure of the low chromosphere and about the existence of dust shells around the Sun; these, too, would be reexamined.

HOW IS THE CORONA HEATED?

As you may remember from chapter 5, the corona is much hotter than the surface of the Sun: one to two million Kelvin even in its quiescient moments. Since heat doesn't run "uphill," from cold to hot places, some special mechanism must maintain the corona's high temperature. In 1991 several mechanisms were hypothesized for this task.

One idea, going back a long way (see chapter 5) was that Alfven waves are somehow generated in the photosphere and ride up along magnetic-field lines into the corona, where they dump their energy as heat, like an ocean wave breaking on a beach. An Alfven wave is a transverse wave, similar to the traveling wave you produce when you snap a whip. In the corona, the "whip" is a bundle of magnetic-field lines, and the decrease of plasma density with increasing height corresponds to the tapering thickness of the whip.

The idea of wave heating makes physical sense, but even after twenty-five years of searching, nobody had ever detected Alfven waves unambiguously. One good reason is that the corona is transparent to visible light and also to X rays, so that each line of sight at the solar limb samples a very long path through the corona. The observer thus records an average of the corona's motions, temperature, or density. Even if Alfven waves exist, they might not show up as local brightenings or periodic motions because of such averaging. But observers continued to hope that if an isolated coronal loop could be observed during an eclipse, it might reveal the long-sought waves.

Eugene Parker, an eminent and outspoken theorist at the University of Chicago, was unimpressed with this possibility. He had strong theoretical objections to Alfven-wave heating. Such waves would be difficult to generate in the photosphere, he said, and even more difficult to dissipate in the corona. He proposed instead that tiny flares occur constantly in the corona. Astronomers had realized in the 1970s that the footpoints of coronal loops are anchored in the photosphere and that convective motions shuffle these footpoints of the field in a random walk. The coronal loops are twisted and braided as this happens. In 1972, Parker showed that a twisted coronal loop would be unstable unless the twist was uniform along the loop. Since that is a very unlikely situation in the real corona, he concluded that the excess energy, stored in the twisted field, would rapidly dissipate as tiny flares.

Parker revived these ideas in the late 1980s and presented a stream of arguments to bolster the concept. Some X-ray and extreme-ultraviolet observations made around that time could be interpreted as evidence for this "nanoflare" hypothesis. Parker estimated their energy could be a hundred or a thousand times smaller than that of the smallest flares yet observed. But with a big telescope, a dark sky, and a long eclipse, they might just be detected.

According to one of Maxwell's equations of electromagnetism, a twist in a magnetic field is equivalent to an electric current that flows along the field. This raises the possibility of another heating mechanism: heating by dissipation of coronal electric currents. (See Rosner's suggestion in chapter 5.) The currents would lose their energy because of the resistance of the plasma and, unlike Parker's nanoflares, would heat the corona evenly and continuously, like the coils in your toaster. This gentle type of heating would be virtually impossible to confirm by any conceivable observation.

In 1991, the stage was set for a decision among these competing ideas. Each mechanism had its proponents. Jay Pasachoff, a professor at Williams College and a veteran eclipse observer, was ready to search once again with his special instrument for the elusive Alfven waves. He decided for several reasons to set up at sea level on the west coast of the island. Barry LaBonte, from the University of Hawaii, planned to hunt the tiny nanoflares with the University's 2.2-meter telescope atop Mauna Kea. Serge Koutchmy, from the Paris Institute of Astrophysics, would mount an all-out effort to detect Alfven waves with the 3.6-meter Canada-France-Hawaii telescope.

Pasachoff had tried twice before to find Alfven waves. He was guided in his search by predictions made by theorists Jim Ionson and Joe Hollweg that the most likely wave periods lie below ten seconds. Pasachoff designed an instrument to look for short-period brightness variations in coronal loops. His technique was to form an image of the corona in the bright-green emission line, emitted by Fe XIV, place thin optical fibers at suitable places in the image during totality, and record the brightness fluctuations photoelectrically at a high sampling rate. Adjusting the fibers in mid-eclipse takes a cool head and a steady hand, as you can well imagine!

At the 1980 eclipse in India, Pasachoff and Don Landman recorded brightness fluctuations about 1% above the noise level in the range of 0.5 to 2 seconds. These results were encouraging, but not conclusive, since a number of factors (sky-transparency fluctuations, for example) could have contaminated the data. So for the 1983 eclipse in Java, Pasachoff added a separate sky-background channel in order to eliminate the principle contaminant. Once again he found 1% excess power around 1 second periods, with much greater certainty.

However, his evidence for Alfven waves was criticized by other observers and, on theoretical grounds, by Parker, who was skeptical that Alfven waves as short as 1 second could be excited in the photosphere or could dissipate sufficiently rapidly to heat the corona. Pasachoff, he said, might have detected standing waves along a loop (like those on a guitar string), not propagating waves. But in that case, the power involved would be insufficient to heat the corona. Pasachoff decided to try once again, with improved equipment, to convince the skeptics.

Meanwhile, Koutchmy and his French associates set up at the huge 3.6-meter telescope. Koutchmy had detected 43 second period doppler shifts in the corona in 1983 without an eclipse, and he was determined to find out this time if wave heating was detectable. He mounted a 70 millimeter stop-motion camera at the telescope's prime focus and planned to film the heating process in white light. An auxiliary video camera would record images at 30 frames a second, faster than the film camera.

LaBonte and his collaborators (including the author) reasoned that heating by nanoflares might best be demonstrated by making digital movies of the corona simultaneously in three well-known emission lines, each sampling a different temperature range. The yellow, green, and red lines they chose are formed respectively near 3, 2, and 1 million Kelvin. LaBonte's idea was to look for tiny impulsive "sparks" that first appeared hot and gradually cooled. The frequency and energy of these events were unknown, but hopefully covered an observable range. He could reasonably expect, from what is known about ordinary flares, that the weaker nanoflares occur far more frequently than the stronger ones. Even if LaBonte could not actually detect nanoflares, he would be able to fix better limits on their energy than anyone previously.

These three investigators and their associates mounted their equipment and awaited the morning of July 11.

HOW DENSE IS THE CORONA?

Take a look at Figure 12.3. The brightness at any point in such a photograph is a direct measure of the total amount of gas along the line of sight through that point, regardless of how the gas is distributed along the line. In isolated plumes or loops, therefore, the brightness is a measure of the actual gas density. So the sharper the image,

Fig. 12.3. The solar corona, as recorded by one of the cameras of the Multistation International Coronal Experiment (MICE). This image was obtained in Brazil.

and the more isolated the structure being measured, the more accurate the estimate of the density.

Koutchmy had noticed this effect: that the density one infers from an image depends on the sharpness of the image and whether one could locate individual structures. The coronal densities we call typical, he claimed, are far off the mark: they are just averages taken from images in which many sharp features overlap in the line of sight. What would be the density of a truly isolated coronal loop? He already had some idea from his observations at previous eclipses that the true densities were factors of two and three times larger than the averages. If so, that could have an important bearing on the amount of power required to heat these loops. He determined to try to confirm these old results with his film and video cameras at the prime focus of the CFH telescope.

Leon Golub, a solar astronomer from the Smithsonian Astrophysical Observatory had a similar goal. He was preparing to fly a radically new X-ray camera aboard a rocket during the eclipse. As you can see in Figure 12.4, the hot corona stands out against the cool dark

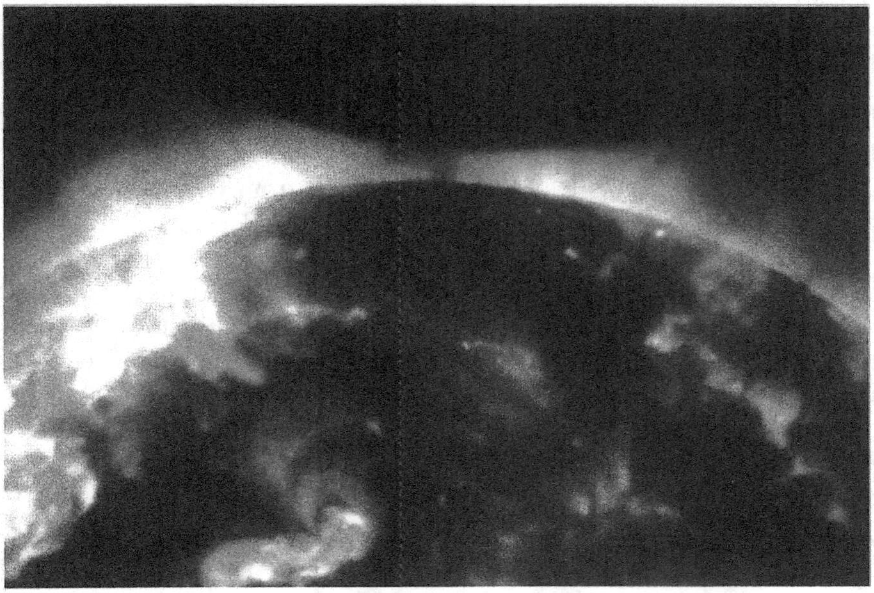

Fig. 12.4. The solar corona, photographed in X rays by the Normal Incidence X-ray Telescope of the Smithsonian Astrophysical Observatory.

photosphere in an X-ray picture of the Sun. Golub's X-ray images, obtained on test flights, were four or five times sharper than any taken before. Figure 12.4 shows narrow loops only 300 kilometers in diameter—a factor of five smaller than any seen in a white-light photograph like Figure 12.3. With that kind of spatial resolution and with much less confusion of structures along a line of sight, Golub had been able to derive more reliable plasma densities, and these were several times higher than earlier estimates. But were they typical?

Golub was exploiting a new technolgy in X-ray astronomy: normal incidence X-ray mirrors. Until the 1990s, X-ray telescopes used mirrors that could only reflect rays that approached the mirror at grazing incidence angles. These telescopes were complicated, expensive, and fragile. They produced solar images with a resolution no better than 2 arcseconds (1500 km). Golub's mirrors use special coatings, similar in principle to the antireflection coatings on an ordinary camera lens, but designed to work at wavelengths near 40 angstroms, where a number of coronal iron lines cluster. Such mirrors reflect rays incident nearly perpendicular to their surfaces, just as optical telescopes do. They are cheaper, sturdier, and smaller. They have produced superb photographs of two million degree coronal loops, similar to Figure 12.4.

Golub would launch his rocket experiment from the White Sands Missile Range in New Mexico, not in Hawaii. The eclipse would only be partial there, but that suited his purpose perfectly, since he wanted to photograph the X-ray corona against the solar disk, not only at the limb. His flight would last about five minutes and was timed to start simultaneously with second contact in Hawaii. Golub was hoping for coronal images that would perfectly complement those taken at Mauna Kea. Figure 12.6 (below) shows some of his results, to which we'll return shortly.

HOW DOES THE CHROMOSPHERE'S TEMPERATURE VARY?

As we read in chapter 10, the chromosphere is a region of intermediate temperature between the two-million-degree corona and the five-thousand-degree photosphere. Figure 4.4, an image photographed in the hydrogen alpha line, shows that the chromosphere is not a smooth shell but concentrates mainly on the boundaries of supergranule cells to form a "chromospheric network." At high magnification, the network resolves into spicules, the needle-shaped structures that probably outline magnetic-flux tubes. The spicules are so narrow (less than 700 kilometers in diameter) that their internal temperature distribution is difficult to study.

One way to beat this problem is to measure the spicule's infrared radiation ("heat"). At wavelengths longer than 10 microns the spicules are opaque to infrared light, so that a measurement of intensity yields a direct measure of their temperature. By measuring the intensity at several wavelengths, one gets a number of independent checks on the temperature. Of course, this method only samples the cool portions of spicules, but that is sufficient to constrain spicule models.

Charles Lindsey and his coworkers at the University of Hawaii had carried out just such an experiment during the 1983 eclipse aboard NASA's Kuiper Airborne Observatory. They measured the intensity profile of the chromosphere at the solar limb at 30, 50, 100, and 200 microns and found that the chromosphere extended far higher than contemporary models predicted. They attributed this result to the presence of spicules, which the models did not include. Now, in 1991, they had a chance to use the occulting edge of the Moon during a total eclipse to gain far higher spatial resolution. They set up their detector at the 15-meter James Clark Maxwell telescope and prepared to observe the spicules at a wavelength of 1.3 millimeters. Measurements at this long wavelength would be particularly sensitive to the spicule temperature.

Meanwhile, Drake Deming and Don Jennings from NASA were installing "Celeste," a cryogenically cooled spectrometer, at the focus of NASA's 3-meter Infra-red Telescope Facility. To protect their telescope mirror from direct sunlight, they covered it with a huge sheet of polypropylene, which transmits the far infrared but scatters visible light. They were planning to observe the solar limb at 12.32 microns. Back in 1960, James Brault and Larry Testerman had discovered a very strange line in the photospheric spectrum at this wavelength. Unlike all other photospheric lines, this one was an emission line, slightly brighter than the photosphere (see Figure 12.5). Brault and Testerman thought originally it was a glitch in their observations. The line was later identified as due to neutral magnesium. Because of its long wavelength, it splits by large amounts even in weak magnetic fields and is therefore very useful to solar physicists. Before it can be used as a tool, however, one needs to know where in the solar atmosphere the line is formed. Theorists predicted it formed in the low chromosphere. Deming and Jennings were about to check that prediction.

Elsewhere on Mauna Kea, Alan Clark from the University of Calgary had a similar goal. He was about to measure the height above the solar limb at which the strong spectral bands of the carbon-monoxide molecule form. These bands appear between 4.4 and 5.4 microns in the spectrum of the photosphere. Since carbon-monoxide molecules break up at temperatures above about 4000 Kelvin, the existence of these bands implies that the photosphere contains gas somewhat cooler than this. But this conclusion contradicts the classical one-dimensional model of the photosphere, which has a minimum temperature of 4400 Kelvin. Clearly, the model would have to be

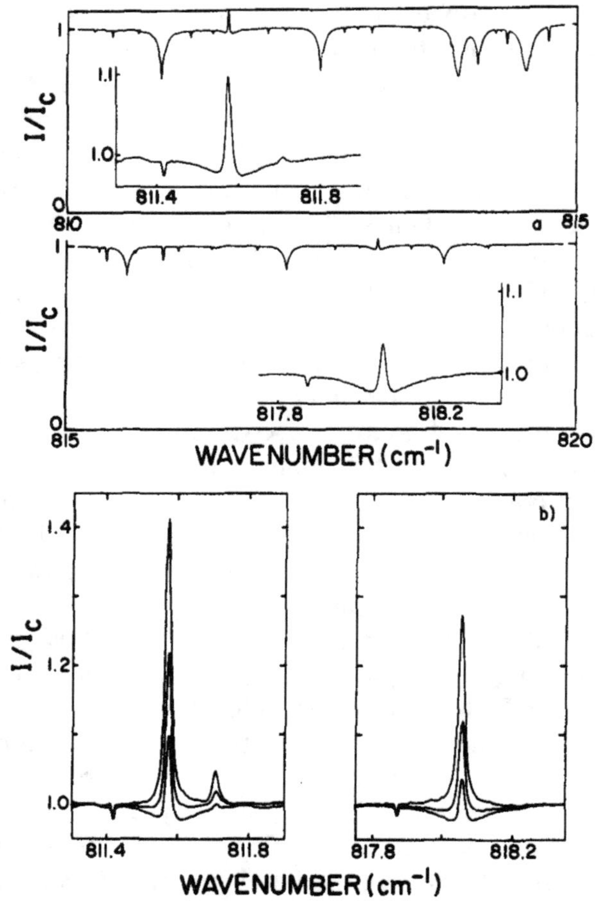

Fig. 12.5. Profiles of the magnesium line at 12 microns. This strange line is in emission even at the center of the solar disk.

modified, and as a first step, Clark would determine the height variation of emission, using the Moon's motion to provide a smooth, accurate height scale.

Two groups of observers on Mauna Kea were not at all interested in the Sun itself. They were there to confirm, once and for all time, the existence of dust shells around the Sun. Theorists had predicted that cometary ice grains would heat up as they fell slowly inward toward the Sun. When they reached a critical distance, they would evaporate and emit infrared light strongly. Contradictory reports of such shells had been published for twenty-five years. Alan Peterson of the University of New Mexico claimed he detected one at a distance of 4 solar radii from the Sun's center at the 1966 eclipse and again in 1973. Robert MacQueen of the High Altitude Observatory had discovered three discrete shells, at 3.5, 8.7, and 9.2 solar radii, using a balloon-borne coronagraph without an eclipse. But other observers had failed to confirm these findings. It was time to make definitive measurements, and Mauna Kea was the ideal infrared site for the purpose. And so, MacQueen had come with his collaborators and, elsewhere on the volcano, Pierre Lamy of the Marseille Observatory was there with his group.

HOW RAPIDLY DOES THE OUTER CORONA CHANGE SHAPE?

The solar corona is constantly changing shape because its magnetic field is distorted by motions that originate in the photosphere. Some shape changes are catastrophic: a whole streamer can rip out of the corona and blow into space. (See Figure 12.6.) Such ejections occur once or twice a day. Slow smooth changes also occur, but without an eclipse, these can be observed only in the inner corona. We know very little about the motions of the outer corona, beyond, say, two solar radii.

One way to investigate the outer corona is to station identical telescopes along the path of a total eclipse. At any fixed location on the path, totality lasts a few minutes at most, but the eclipse can be seen from different locations at times that differ by as much as several hours. By taking observations at points spaced along the path, a time-lapse record of the outer corona can be constructed.

Serge Koutchmy organized such a chain of telescopes for the total eclipse of 1972 and again in 1973 but was thwarted first by bad weather and later by equipment failure. Koutchmy was one of the first to employ special filters at the prime foci of his telescopes to flatten the very steep gradient of coronal brightness along an outward radius. This technique enabled him to photograph the fine-structures in both the inner and outer corona simultaneously.

For the July eclipse, Koutchmy and I assembled a dozen astronomers from seven countries: Russia, Ukraine, Slovakia, Belgium,

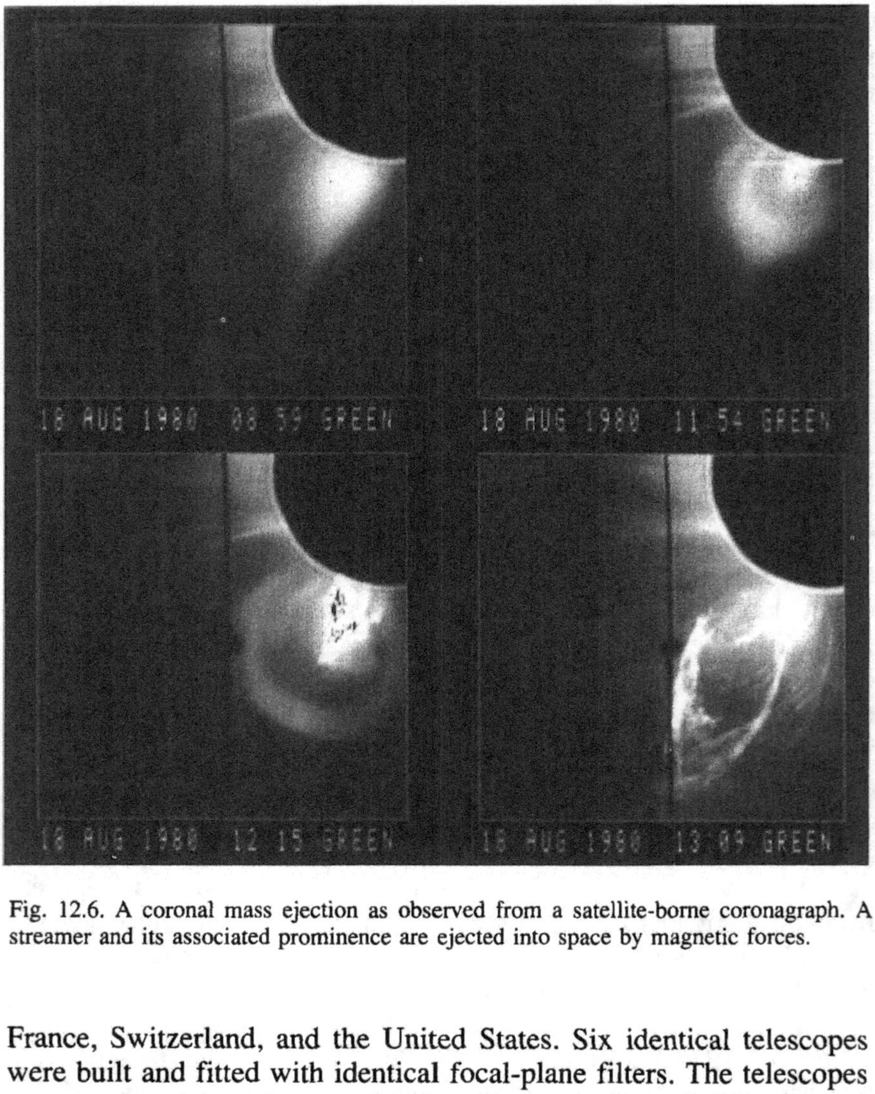

Fig. 12.6. A coronal mass ejection as observed from a satellite-borne coronagraph. A streamer and its associated prominence are ejected into space by magnetic forces.

France, Switzerland, and the United States. Six identical telescopes were built and fitted with identical focal-plane filters. The telescopes were stationed in pairs: two in Hawaii, two in Baja California, and two at a remote village in central Brazil. The time difference of the eclipse between Hawaii and Brazil amounted to three and a half hours—sufficient time to expect observable changes in the outer corona. As the leader of this consortium, I had the privilege of inventing a suitable name. I chose the "Multistation International Coronal Experiment," or "MICE."

THE CLIMAX: TOTALITY!

On July 11 the observers were on the summit at dawn. The eastern sky was beautiful, admittedly, but discouraging. An intertropical front had arrived during the night and now the pink sky was laced with two layers of cloud: high cirrus and low cumulus. Moreover, ten days earlier, the dust cloud from a volcanic eruption of Mount Pinatubo in the Philippines had reached Hawaii and was covering the sky with a thin veil. Even worse, a layer of fog blanketed the island just a few hundred feet below the summit. The prospects were dismal, but having come this far and worked this long, the scientists had no choice but to go through their checkout routines and hope for a change. Perhaps the legendary clearing of clouds that often accompanies a total eclipse would save the day.

Drake Deming and Don Jennings had worked all night and the previous day and were near exhaustion. They had had stubborn problems with their detectors and had to jury-rig a substitute. An hour before totality, they had their first chance to check their signals with sunlight. To their delight the system worked. Now if only the clouds would cooperate . . .

While the sun climbed slowly toward a break in the clouds, the fog layer began to rise ominously toward the summit. The astronomers joked with the magazine photographers who were taking snapshots of the pretty clouds. "Enjoy yourselves, you guys! Maybe cloud pictures is all anybody will get!"

As the sky gradually dimmed, past first contact, a light breeze picked up, pushing the fog toward the summit. Now there was a real danger of dew forming on the telescope mirrors. But, no doubt about it, the sky was beginning to clear, just as the old folktale said it would. It was a race now between the fog and the sun. If the sun could reach that break in the clouds, the observers had a slim chance.

As second contact approached, the astronomers began their final routines and prepared to open their telescope domes. Second contact was 7 seconds away, 6, 5, 4, 3, 2, 1, CONTACT! The photographers outside the domes cheered, but the astronomers inside were too busy to notice. There, by a miracle, was the corona, shimmering in all its glory. The cameras snapped, the tape recorders hummed, and the astronomers murmured their thanks to whatever deity governs the Hawaiian Islands these days. Four minutes later, totality had passed.

The telescopes were stowed and the cameras turned off. The astronomers sat back, exhausted, grateful, and enormously pleased. The television crew of *Nova* caught a candid shot of Serge Koutchmy celebrating with champagne. It had been a close race. A few minutes after totality the fog covered the summit.

RESULTS

Within a year, the results of the eclipse experiments began to appear in the journals. Here is a brief summary of what was found.

None of the attempts to identify the cause of coronal heating succeeded. Pasachoff was clouded out, LaBonte lost some crucial data, and Koutchmy's videos showed no wave motion. But Koutchmy's video camera did record an extraordinary and totally unexpected event: the ejection of a tiny blob of gas (only 1000 kilometers in diameter) into the corona. Some theorists had proposed this type of event as the way mass is fed into the solar wind.

Both Koutchmy and Golub recorded fine-structures in the corona and derived plasma densities three to five times higher than previous estimates, based on coarser spatial resolution. Lindsey found, once again, evidence of cool gas (only 8000 Kelvin) in spicules, extending at least 3000 kilometers above the limb of the sun. New models of the low chromosphere would have to incorporate these results.

Deming and Jennings established that the 12 micron line is formed below the temperature minimum, in accord with the best models, but extending much higher than the models allow. Clark pinpointed the height at which the carbon-monoxide bands form: 125 kilometers above the limb.

Despite the veil of volcanic dust over the sky, both groups searching for interplanetary dust shells agreed that none were detectable at this eclipse. This leaves open the question of what earlier observers saw. Are dust shells transient events in interplanetary space?

Finally, the MICE team obtained several superb white-light images of the corona. (See Fig 12.3.) Two of them, taken three hours apart, showed that a cluster of long plumes on the west limb had untwisted during the eclipse, implying that magnetic waves must propagate in the outer corona at speeds exceeding 1000 km/s.

In the three years since the Hawaiian eclipse, more detailed analyses of the data have appeared and new questions have arisen. Future eclipses will certainly help to answer them.

Postscript

As this book goes to press, scientists are analyzing the observations they obtained at the total eclipse of November 3, 1994. The path of totality passed over northern Chile, Bolivia, Brazil, and Argentina and then out over the Atlantic. Most astronomers chose to observe the eclipse in Putre, Chile, one of the driest cloud-free regions in the world. As luck would have it, the sky was covered with light cirrus that day; nevertheless, we fielded some interesting experiments and obtained useful data during three minutes of totality.

Among the more novel experiments was a search for infra-red coronal emission lines by J. Kuhn and Matt Penn of the National Solar Observatory. The infra-red spectrum is virtually unknown, because the digital cameras needed to explore it have become available only recently. The spectrum may contain some lines that would be particularly valuable for measurements of coronal density, temperature, and magnetic field. Kuhn and Penn searched the spectrum from 1.0 to 1.5 microns and easily found the two known lines of Fe XIII and a Silicon X line. It will take careful calibration and analysis to reveal any other lines.

J. Pasachoff, an intrepid eclipse observer, applied modern technology to an old technique for measuring the average temperature of the corona (see pages 93–94). Several experiments aimed at measuring the line of sight motions in the inner corona: Ray Smartt (National Solar Observatory) fielded an improved version of his Fabry-Perot interferometer; I. Kim (Moscow State University) had a similar rig in Brazil; and R. Cowsik from Bangalore, India used a multi-slit spectrograph for the same purpose. S. Koutchmy brought a large French team, which operated a fast spectrograph to map velocities in coronal streamers. These spectroscopic experiments are less sensitive to cloudy skies than imaging devices and should have done well.

Images of the white-light corona were obtained by O. R. White of the High Altitude Observatory, by Koutchmy, and by Kim and her associates. The author attempted to record the green coronal emission line far out in streamers, but was foiled by the cirrus clouds.

Perhaps the most daring experiment was made by a group of French amateurs who scaled Cerro Parinacota, a 6,342-meter volcano, to photograph the eclipse. They were successful and have published their snapshot already. They may very well start a new fad!

Where do we go from here? There are only four total eclipses left this century:

Date	Period of Totality		Place
Oct. 24, 1995	2 min	10 sec	India, Thailand, Cambodia
Mar. 9, 1997	2	51	Indonesia
Feb. 26, 1998	4	9	Colombia, Venezuela, Aruba
Aug. 11, 1999	2	23	Europe

The 1998 and 1999 eclipses will be well attended. See you there!

Glossary

Air-glow: Radiation of excited atoms in the Earth's atmosphere.

Angstrom: 10^{-8} cm.

Angular momentum: The tendency of a spinning body to maintain its motion, measured by mass × (radius arm)^2 × angular speed.

Apogee: The point in the Moon's orbit farthest from the Earth.

Arcminute: 1/60 of a degree.

Arcsecond: 1/360 of a degree.

Atomic weight: The number of protons plus neutrons in an atom.

Chromosphere: A dynamic inhomogeneous layer between the corona and photosphere.

Comparator: An instrument for accurate measurement of small distances.

Contacts: The instants in a solar eclipse when the Moon's disk "touches" the sun's disk.

Continuous spectrum (continuum): A spectrum without spectrum lines in emission or absorption.

Corona: The hot outer atmosphere of the Sun.

Coronagraph: A special telescope that produces artificial eclipses of the Sun by masking its image with an opaque disk.

Coronal condensation: A coronal region of unusually high brightness and density.

Coronal hole: A coronal region of unusually low brightness (density).

Coronal plumes: Fine, thread-like structures.

Coronal streamer: A fan-like structure, extending at least 2.0 solar radii from the Sun.

Day, mean solar: The average interval between successive crossings of a local meridian by the Sun.

Doppler shift: A displacement in frequency of radiation, due to the source's motion.

Eccentricity of an orbit: The ratio of the distance between the foci of an orbit and the major axis. A measure of the "flatness" of the orbit's ellipse.

Eclipse of Sun—annular: An event during which the Moon leaves exposed a bright ring at the Sun's edge.

Eclipse of Sun—central: Total or annular eclipse.

Eclipse of Sun—partial: An event during which the Moon covers part of the Sun.

Eclipse of Sun—total: An event during which the Moon completely covers the Sun.

Ecliptic limit: The maximum angular distance of the Sun from a node that permits a solar eclipse.

Electron-volt: A unit of energy. A kilowatt-hour equals 2×10^{25} electron volts.

Equinoxes: The points of intersection (on the sky) of the ecliptic and celestial equator (the extension to the sky's "sphere" of the Earth's equator.

Erg: A unit of energy. A kilowatt hour equals 4×10^{13} ergs.

Faraday rotation: The turning of the plane of polarization of radiation within a gas that contains a magnetic field.

Fe XIII, (etc.): Iron atom with 12 electrons missing (and so on).

Field (magnetic, gravitational): A modification of empty space that generates a force on a moving charge, mass, or magnet.

Fraunhofer lines: Absorption (dark) spectrum lines in the solar spectrum

Geomagnetic storm: A disturbance of the Earth's magnetic field.

Grating: An optical element (flat or curved) on which many grooves have been cut into a thin reflective layer.

Hydrostatic equilibrium: A static condition in which the pressure at each point of a fluid or gas supports the weight of overlying layers.

Infra-red: The portion of the electromagnetic spectrum with wavelengths between about 1 and 300 microns.

Interferometer: An instrument that combines radiation from a source, in several directions, and measures either the size of the source or its spectrum.

Ion: An atom with a positive or negative electric charge.

Ionization: The process of removing one or more bound electrons from an atom.

Kilometer: 0.621 miles.

Kinetic energy: The energy of a body related to its motion.

Limb-brightening: A brightening at the edge of the apparent solar disk.

Limb: The edge of the apparent disk of a planet, Moon, or Sun.

Magnetic field line: The direction, in space, of the field that interacts with charged particles.

Magnitude (stellar): The brightness of a star. A difference of one magnitude corresponds to a factor of 2.51 in brightness.

Mass flux: The amount of material (grams) passing through unit area (cm^2) in unit time (second).

Meridian: The vertical great circle that passes through the zenith and the north (south) point on the horizon.

Microdensitometer: An instrument that measures the transparency of photographic plates or films.

Momentum: The tendency of a body to continue its motion; equal to mass × velocity.

Month, draconic: The interval (27.21 days) between successive passes of the Moon past a node.

Month, sidereal: The interval (27.32 days) between successive passes of the Moon past a fixed star.

Month, synodic: The interval (29.53 days) between successive new moons.

Nebula: A body of interstellar gas, illuminated by nearby stars.

Node: The projection on the sky of the intersection of the Sun's and Moon's path.

Node, ascending: The node at which the Moon crosses from south to north of the ecliptic.

Node, descending: The node at which the Moon crosses from north to south of the ecliptic.

Objective (lens): The main optical element in an instrument.

Penumbra: The partially-lit part of the Moon's shadow in a solar eclipse.

Perigee: The point in the Moon's orbit nearest the Earth.

Periodic table: A systematic arrangement of the elements according to their atomic number or weight. The arrangement emphasizes common elemental properties.

Phase (of an oscillation): A particular instant in a recurring event (i.e. beginning, maximum, etc.).

Phases of Moon: The partially-lit appearance of the Moon at different points in its orbit.

Photon: An elementary quantity of light, that acts like a particle in collisions with atoms.

Photosphere: The visible "surface" of the Sun.

Plasma: An ionized gas containing equal numbers of negative and positive electrical charges.

Polarized light: Light with a preferred direction of vibration of the electromagnetic field.

Poynting-Robertson Effect: A dust particle in orbit about the Sun experiences a retarding tangential force that is proportional to the radial force of radiation pressure, and proportional to tangential speed of the dust particle.

Prominence: A flat wall-like mass of cool gas, suspended in the corona.

Radiosonde: A package of instruments, equipped with a radio transmitter, that is carried aloft in a balloon or rocket.

Revolution: The turning of a body about an axis that is within another body.

Rotation: The turning of a body about an axis within itself.

Saros cycle: A period of 18 years in which eclipses tend to recur.

Spectrogram: The photograph of a spectrum.

Spectroscope: A device for viewing the spectrum of a sample element.

Spectrum, line: Atomic radiation of almost a single wavelength.

Spectrum: A mixture of radiation with many wavelengths.

Total electron content: The total number of free electrons in a vertical column, 1 cm^2 in area.

Turning point of eclipse path: The point furthest north or south.

Ultra-violet: The portion of the electromagnetic spectrum with wavelengths between about .1 and .3 microns.

Umbra: The darker part of the Moon's shadow in a solar eclipse.

Universal Time: The local time at Greenwich, England.

Wavelength: The characteristic length (say from one maximum to the next) of a wave.

Wind, solar: The stream of gas emitted continuously by the Sun into interplanetary space.

X-ray: The portion of the electromagnetic spectrum with wavelengths between about .01 and 100 angstroms.

Year, eclipse: The interval (346.62 mean solar days) between successive passes of the Sun past a lunar node,

Year, sidereal: The interval (365.25636 mean solar days) between successive passes of the Sun past a fixed star.

Zenith: The overhead direction.

Zodiacal light: Sunlight visible in the ecliptic, scattered from interplanetary dust.

Index